Springer Texts in Statistics

Advisors:
Stephen Fienberg   Ingram Olkin

William S. Peters

Counting for Something
Statistical Principles and Personalities

With 46 Illustrations

Springer-Verlag
New York Berlin Heidelberg
London Paris Tokyo

William S. Peters
Robert O. Anderson Schools of Management
University of New Mexico
Albuquerque, NM 87131
U.S.A.

AMS Classification: 62-01

Library of Congress Cataloging in Publication Data
Peters, William Stanley,
    Counting for something.
    (Springer texts in statistics)
    Bibliography: p.
    Includes index.
    1. Statistics—History. I. Title. II. Series.
QA276.15.P47   1986   519.5'09   86-11866

Typeset by Asco Trade Typesetting Ltd., Hong Kong.
Printed and bound by R. R. Donnelley & Sons, Harrisonburg, Virginia.
Printed in the United States of America.

9 8 7 6 5 4 3 2 1

ISBN 0-387-96364-2 Springer-Verlag New York Berlin Heidelberg
ISBN 3-540-96364-2 Springer-Verlag Berlin Heidelberg New York

Dedicated to the memory of

E. Douglass Burdick
1905–1961

and

J. Parker Bursk
1899–1963

J. Parker Bursk and E. Douglass Burdick introduced me to the subject of statistics at the Wharton School of the University of Pennsylvania. Parker Bursk was Chairman of the Department of Economic and Social Statistics at the Wharton School from its founding in 1932 until his death. The department at Wharton may well have been the first statistics department in a collegiate business school. In addition to his teaching, he served the Pennsylvania Department of Labor and Industry, the Wage Administration Agency of the War Department, and the United States Air Force in statistical and administrative roles.

Doug Burdick taught statistics at the University of Pennsylvania and the Wharton School from 1930 to 1961. In addition, he directed a number of statistical studies in housing, medicine, and public health. He undertook assignments for the International Cooperation Administration in Egypt, India, and Turkey.

Parker Bursk and Doug Burdick were extraordinary teachers in quite distinctive ways. Parker was thorough and incisive, and Doug was more iconoclastic and inspirational. What they had in common was a dedication to learning, to their students, and to human values. In my life, they counted for something.

Contents

List of Tables

List of Figures

CHAPTER 1

Statistics—The Word

Two views about quantification are expressed by a scientist and a poet. First, Lord Kelvin:

> When you can measure what you are speaking about, and express it in numbers, then you know something about it; but when you cannot measure it, when you cannot express it in numbers, your knowledge is of a meager and unsatisfactory kind.

And now e. e. cummings:

> While you and I have lips and voices which
> are kissing and to sing with
> who cares if some one-eyed son-of-a-bitch
> invents an instrument to measure spring with.

One can agree about the limitations of measurement, but at the same time observe that modern society is becoming more and more measurement conscious. Are we becoming at once more sophisticated and more superficial?

This book illustrates the principles of quantification, especially as applied to economics and social science. It draws liberally on the history of the ideas and personalities that have been prominent in the development of quantitative applications. The book is more a "show-and-tell" account than a "how-to-do-it" manual. The main ideas were advanced by people who spent their professional lifetimes counting and measuring—or thinking about ways that counting and measuring could be improved.

Descriptive Statistics

The field of knowledge that deals with counting and measuring is known as "statistics." The word "statistics" is also used to describe the results of counting and measuring—that is, the numbers that counting and measuring produce. Winston Churchill, when still First Lord of the Admiralty in 1939, refers to both uses of "statistics" in a request for information.

> Surely the account you give of all these various disconnected statistical branches constitutes the case for a central body which should grip together all admiralty statistics and present them to me in a form increasingly simplified and graphic. I want to know at the end of each week everything we have got, all the people we are employing, the progress of all vessels, works of construction, the progress of all munitions affecting us, the state of our merchant tonnage, together with losses, and numbers of every branch of the Royal Navy and Marines. The whole should be presented in a small book such as was kept for me when Sir Walter Layton was my statistical officer at the Ministry of Munitions in 1917 and 1918. Every week I had this book, which showed the past and weekly progress, and also drew attention to what was lagging. In an hour or two I was able to cover the whole ground, as I knew exactly what to look for and where. How do you propose this want of mine should be met? [1]

In this request, Churchill wants more than just numbers. He wants numbers organized so as to convey *information*. Only then will they aid understanding and provide a guide to action.

The art of organizing numerical information is often called *descriptive* statistics. For example, development of the system of national income statistics by the U.S. Department of Commerce in the 1930s was a major accomplishment in descriptive statistics. A concept of income had to be developed and refined, and economic activities classified according to categories related to this concept. Ways of counting or measuring activity in the various categories had to be found, and the whole organized into a systematic set of accounts designed for periodic reporting.

Such a system was not in hand at the time of the Great Depression of the 1930s. It was as if the economic and political commanders of that time did not have a summary "book" of the kind that Churchill had in 1917–1918 and again during World War II for directing military activities.

Other examples of descriptive statistics systems can be found in education, business, and sports. More and more they are called *information systems*, and they are organized for data storage and retrieval by computers. The purpose of the system is to define and monitor an ongoing activity to provide a guide for action.

Inferential Statistics

When data or information is based on a sample of possible cases there is a need for generalizing from the particular cases studied to the universe of possible cases. The process is called statistical inference and the methods developed for the purpose are called inferential statistics.

Two clear examples where inferential statistics are needed are in generalizing the results of election polls and in using sampling to check the quality of industrial materials. Forecasts of presidential elections in the United States are made from samples of 2500 persons in the entire nation. Checks on the quality of incoming materials or outgoing product in a manufacturing operation are typically based on a correspondingly small fraction of the total product. How can the results of such a small sample be projected to the large population involved?

The foundations of statistical inference lie in probability, and through probability statistics is connected with mathematics. We find that early contributors to statistical theory were mathematicians. Pierre Simon, the Marquis de Laplace (1749–1827), one of the great contributors to probability theory, indicates that a question put by the Chevalier de Mere to the mathematician Blaise Pascal (1623–1662) about a gambling problem "caused the invention of the calculus of probabilities." [2]

Carl Friedrich Gauss (1777–1855) first applied the theory of probability to the investigation of errors of observation, using techniques based on the work of Laplace. The data of interest to Gauss were astronomic observations, such as repeated measurements of the distance from the earth to the sun. Such measures would be determined in large part by the "true" distance, but a multiplicity of causes would produce errors of observation in individual measures. In studying the character of such errors, Gauss was investigating the same principles that determine how the percent of voters favoring Candidate A in a sample varies from the true percentage favoring the candidate among all voters in the nation. The sample percentage is determined in large part by the true percentage for the population, but also varies from it owing to a multiplicity of causes that we call chance. There are errors of measurement in using the sample percentage as an estimate of the population (national) percentage.

Richard von Mises (1883–1953) defined statistics as "the study of sequences of figures derived from the observation and counting of certain repetitive events in human life." [3] The appearance of a defective product in a stream of output from a manufacturing process is a case in point. As with astronomical observations and voter preferences there is a fundamental cause at work. Here it is the true process capability, or long-run percentage of defectives that the process would produce under the existing conditions. When an essentially constant set of causes prevails, the set or sequence of observations can be termed a *collective*. Figures like the percentage of defective products in the stream then tend to limiting long-run values which we call probabilities. These values, or probabilities, are not affected by any system of selection, such as time-of-day, machine operator, and so on. This condition von Mises calls the *principle of randomness* or the *principle of the impossibility of a gambling system*. This principle is as fundamental to probability as the law of conservation of energy is to mechanics. The impossibility of a gambling system can be compared to the impossibility of creating a perpetual motion machine.

Adolph Quetelet (1796–1874) was one of the first to extend the principles of measurement errors to the social and political spheres. A Belgian, Quetelet was the originator of the phrase l'homme moyen, or the average man. Quetelet was

> at once a mathematician, astronomer, anthropometrist, supervisor of official statistics for his country, prime mover in the organization of a central commission on statistics, instigator of the first national census, ardent collector of statistics, university teacher, author of many books and papers, carrying on an extensive correspondence with most of the scholars of his day, and in his leisure hours a poet and a writer of operas. [4]

Among Quetelet's studies was that of crime in society. He was impressed by the regularity of crimes from year to year—by the suggestion in his data of a constant cause system. He commented that

> we pass from one year to another with the sad perspective of seeing the same crimes produced in the same order and calling down the same punishments in the same proportions. Sad condition of humanity! [5]

Let us look at one of Quetelet's data sets from the field of anthropometry. Anthropometry refers to the measurement of mankind. The data in Table 1-1 are the chest measurements of 5738 Scottish soldiers. [6] Here we see clearly that the typical chest measurement is 39 or 40 inches, and that the frequency of measures becomes smaller as one considers chest measurements deviating

Table 1-1. Chest Measurements of 5378 Scottish Soldiers.

Chest measurement (inches)	Number of men	Proportion observed	Proportion from theory
33	3	0.0005	0.0007
34	18	0.0031	0.0029
35	81	0.0141	0.0110
36	185	0.0322	0.0323
37	420	0.0732	0.0732
38	749	0.1305	0.1333
39	1073	0.1867	0.1838
40	1079	0.1882	0.1987
41	934	0.1628	0.1675
42	658	0.1148	0.1096
43	370	0.0645	0.0560
44	92	0.0160	0.0221
45	50	0.0087	0.0069
46	21	0.0038	0.0016
47	4	0.0007	0.0003
48	1	0.0002	0.0001
Total	5738	1.0000	1.0000

from these most common values. Small deviations occur more frequently than large deviations. In the second column the observed frequencies are reduced to proportions or, for the collective involved, estimates of probability. In the final column Quetelet showed the distribution according to the theory of the law of error as developed by Laplace. This law of error was made to serve as a law of distribution of variations among living things. Most of us would view the agreement between the observed proportions and the proportions according to theory as quite close in this example.

Francis Galton (1822–1911) followed in the footsteps of Quetelet and was one of the giants in the development of statistics. Galton was "a brilliant, original, versatile, stimulating scholar" [7] who made important contributions to such diverse areas as composite portraiture, fingerprinting, and the psychology of individual differences. More will be said of him later.

We present here a set of data that Galton collected which contrasts with Quetelet's data shown above. Galton was impressed by differences in the intellectual capacities of men and cited as evidence the differences in scores obtained by those gaining honors in mathematics at Cambridge. [8]

From about 400 students taking degrees annually at Cambridge, some 100 gain honors in mathematics. These 100 are ranked by examination in strict order of merit. About the first 40 are awarded the title of "wrangler," and it is an honor to be even a low wrangler. The highest ranking is called the senior wrangler. The examination lasts for $5\frac{1}{2}$ hours for 8 days. Examiners add up the marks and rank the candidates accordingly.

From a confidential communication, Galton obtained the marks assigned in 2 years, which are presented in Table 1-2. These data do not conform at all to the normal law of error. The symmetry of the chest girth data is lacking here. The typical scores are low in the range of variation, and scores of the top wranglers stretch out toward the upper values. Galton was impressed by the fact that the top wranglers scored twice as many marks as the next highest

Table 1-2. Marks Obtained by Mathematics Honors Students at Cambridge.

Number of marks obtained	Number of candidates	Number of marks obtained	Number of candidates
Under 500	24	4000–4500	2
500–1000	74	4500–5000	1
1000–1500	38	5000–5500	3
1500–2000	21	5500–6000	1
2000–2500	11	6000–6500	0
2500–3000	8	6500–7000	0
3000–3500	11	7000–7500	0
3500–4000	5	7500–8000	1
		Total	200

ranking students, and ten times as many as the most frequently occurring scores.

In the data sets from Quetelet and Galton we have seen two quite different patterns of variability. Karl Pearson (1857–1936) developed a system of measures that defined a number of distinct distribution patterns or curve types. Pearson regarded the collection and classification of data as fundamental to science. But the selection from these of the sequences that should be called natural laws called for disciplined creative imagination. The long road from classification of facts to distillation of a theory is illustrated by Darwin's theory of evolution. [9] Darwin collected data on the voyages of the Beagle from 1831 to 1836. The first outlines of evolutionary theory were published in 1842, but 19 years passed before the final form of the theory emerged.

It would appear that *dealing with variability* is the essential feature of statistics as a method. We have seen this in the examples of both descriptive and inferential statistics that have been given. The chest girths and honors marks collections were arranged to show the pattern of variability in those data. In the case of chest girths a law of error was used to model the variability in the data. The essence of inferential statistics is to generalize from sample data to a larger universe of data. This process comes face-to-face with the problem of differences that can exist between a sample measure and its universe counterpart. In a word, we have to be concerned with the variability of sample measures compared to universe measures.

References

1. Winston Churchill, *The Gathering Storm*, Houghton Mifflin, New York, 1948, p. 70.
2. Pierre Simon de Laplace, *A Philosophical Essay on Probabilities* (translated from the sixth French edition), Dover, New York, 1951, p. 167
3. Richard Von Mises, *Probability, Statistics, and Truth*, Allen and Unwin, London, 1939.
4. Helen M. Walker, *Studies in the History of Statistical Method*, Williams and Wilkins, Baltimore, MD, 1929, p. 31.
5. Ibid., p. 40.
6. Frank H. Hankins, *Adolphe Quetelet as Statistician*, Columbia University Press, New York, 1908, p. 122
7. Edwin G. Boring, *A History of Experimental Psychology*, 2nd ed., Appleton-Century-Crofts, New York, 1950, p. 461.
8. Francis Galton, "Classification of Men According to Their Natural Gifts," in *The World of Mathematics*, Vol. 2 (James R. Newman, ed.) Simon and Schuster, New York, 1956, pp.1173–1189.
9. Karl Pearson, *Grammar of Science*, Dent, London, 1937.

CHAPTER 2

Distributions

The two examples of chest measurements and math scores were presented in the first chapter without much preparation. We hoped it would be more or less obvious how the tables were obtained. We pause now to be more systematic about the concepts of *variables* and *frequency distributions*, and the main measures used to describe them.

Measuring a characteristic of members of a group of persons, objects, or other units gives rise to what statisticians term a variable. Incomes, ages, and heights of individuals are examples of variables. So also are sales volumes, assets, and liabilities of business firms, and tax receipts, expenditures, and bonded indebtedness of municipalities in the United States. One would suppose they are called variables because the measures vary from one unit of observation to the other.

The Frequency Distribution

The frequency distribution of a variable is a list of values of the variable together with the number of occurrences of each value. In the chest measures example each inch in girth and the number of measures nearest that inch were listed, while in the case of the math honors scores ranges of scores, sometimes called classes, were listed along with the numbers of scores in each class.

Whether or not the values of a variable are organized into a frequency distribution, their occurrence over the units under study can be summarized by a few appropriate descriptors. These summary measures are often called *statistics*. So we now have a third meaning of the word. To recap, we have statistics as data, or values of a variable, statistics as methods for studying variables, and statistics as summary measures of variables.

Table 2-1. Ages of Officers Attaining the Rank of Colonel.

47, 47, 48, 48, 48, 48, 49, 49, 49, 49, 49, 49, 50, 50, 50, 50, 50, 50, 50, 50,
50, 50, 50, 50, 50, 51, 51, 51, 51, 51, 51, 51, 51, 51, 51, 51, 52, 52, 52, 52,
52, 52, 52, 52, 52, 52, 52, 52, 52, 52, 52, 52, 52, 52, 52, 52, 52, 52, 52, 52,
52, 52, 52, 52, 53, 53, 53, 53, 53, 53, 53, 53, 53, 53, 53, 53, 53, 53, 54, 54,
54, 54, 54, 54, 54, 54, 54, 54, 54

Let us carry through an example of frequency distribution analysis. Table 2-1 contains data on ages of officers in the Netherlands Air Force who have attained the rank of colonel or better. [1] Already there has been some organization of the data. Its original form might well have been a computer file or an alphabetical list of names with rank and age. With the ages in order we can easily see that the range is from 47 to 54 and the most common age is 52.

Gathering the data into a frequency distribution produces Table 2-2. Now we more easily see the range of variation in ages (47 to 54) and the most typical age (52), and we also see the pattern of variation. The typical age occurs toward the high end of the range, which means that the frequencies stretch out toward the low end of the range. A few senior officers are 3, 4, and 5 years younger than the typical age, but none are that much older.

Table 2-2. Frequency Distribution
of Ages of Senior Officers.

Age	Number of officers
47	2
48	4
49	6
50	13
51	11
52	28
53	14
54	11
Total	89

Summary Measures

The mean and the standard deviation are the summary measures of the distribution of a variable that we want to show now. The mean is a measure of average, or central location, and the standard deviation is a measure of variability.

From the original data of Table 2-1, the mean age is

$$\bar{X} = \frac{\sum X}{n}$$

$$= \frac{47 + 47 + 48 + \cdots + 54 + 54}{89}$$

$$= \frac{4583}{89} = 51.5 \text{ years.}$$

Here, X is a symbol for individual values of the variable, $\sum$ is the symbol for summation, and n is the number of values in the collection. X with an overbar (called X-bar), is the symbol for the mean.

The standard deviation, calculated from Table 2-1, is

$$S = \sqrt{\frac{\sum (X - \bar{X})^2}{n}}.$$

This formula says to find the squared deviation of each value from the mean. Then add these up and divide by the number of values. At this point we have the average of the squared deviations. Then the formula says to take the square root of the average squared deviation. From Table 2-1

$$S = \sqrt{\frac{(47 - 51.5)^2 + (47 - 51.5)^2 + \cdots + (54 - 51.5)^2}{89}}$$

$$= \sqrt{\frac{20.25 + 20.25 + \cdots + 6.25}{89}}$$

$$= \sqrt{\frac{266.25}{89}}$$

$$S = 1.73 \text{ years.}$$

This measure takes into account the deviation (squared) of each value from the mean. The greater the average (squared) deviation of the values from the mean, the greater will be the standard deviation.

If the mean and standard deviation are computed from the frequency distribution listing, we have to proceed a little differently. But we come out in the same place. Table 2-3 shows how the mean would be calculated. Here, we take account of the effect of two 47s, four 48s, etc. on the sum of all the values by mutiplying values times frequency of occurrence. Where X stands for the listed values, the mean now becomes

$$\bar{X} = \frac{\sum fX}{n}$$

$$= \frac{4583}{89}$$

$$= 51.5 \text{ years}$$

as before.

Table 2-3. Calculation of
Mean from Frequency
Listing.

Age X	Number f	fX
47	2	94
48	4	192
49	6	294
50	13	650
51	11	561
52	28	1456
53	14	742
54	11	594
Total	89	4583

The standard deviation calculation proceeds similarly. We find the squared deviation of each listed value from the mean. Then we must take account of the number of values that have the same squared deviations. The calculations are shown in Table 2-4. The standard deviation now becomes

$$S = \sqrt{\frac{\sum f(X - \bar{X})^2}{n}}$$

$$= \sqrt{\frac{266.25}{89}}$$

$$= 1.73 \text{ years.}$$

The result is the same as before.

Table 2-4. Calculation of Standard
Deviation from Frequency Listing.

Age X	Number f	$(X - \bar{X})^2$	$f(X - \bar{X})^2$
47	2	20.24	40.50
48	4	12.25	49.00
49	6	6.25	37.50
50	13	2.25	29.25
51	11	0.25	2.75
52	28	0.25	7.00
53	14	2.25	31.50
54	11	6.25	68.75
Total	89		266.25

Table 2-5. Frequency Distribution of Ages of Lieutenants.

Age	Number of officers	Age	Number of officers	Age	Number of officers
22	3	33	27	44	4
23	10	34	20	45	4
24	18	35	11	46	11
25	23	36	12	47	5
26	33	37	10	48	2
27	19	38	5	49	2
28	23	39	3	50	0
29	18	40	5	51	2
30	15	41	4	52	2
31	23	42	2	Total	348
32	27	43	5		

The listing of single values with their frequency of occurrence will not always produce an effective frequency distribution. For example, the frequency distribution of ages of lieutenants in the Netherlands Air Force, using integer ages, appears in Table 2-5. Using single ages, the distribution is so long as to be unwieldy. Worse, the excessive detail may fail to bring out the underlying distribution pattern. In Table 2-6 we show three different distributions obtained by combining different numbers of single ages.

Which distribution should you prefer as a summary? We feel that the first one shown here still has too many classes and that the third one does not have quite enough. A guide to the number of classes to use is one more than the power of 2 which first exceeds the total number of observations. With $n = 348$ here, the eighth power of 2 is 256, the ninth power of 2 is 512, and this rule of thumb would say use ten classes. The second distribution here has eleven classes, while the third has only seven.

The mean and standard deviation of the lieutenants' ages, found either from the original data or the listing in Table 2-5, are 31.9 years and 6.62 years. That lieutenants are younger than colonels is no surprise and, perhaps, neither is the fact that the lieutenants' ages are much more varied that the colonels' ages. But the figures just given compared with the mean of 51.5 years and standard deviation of 1.73 years for colonels transmit this comparison in a summary fashion.

Graphs

The saying is that "a picture is worth a thousand words." Well executed graphs can quickly convey the essential features of a distribution. In Figure 2-1 the ages of senior officers is shown as a frequency histogram, in which

Table 2-6. Alternative Frequency Distributions for Ages of Lieutenants.

Age	Number of officers	Age	Number of officers	Age	Number of officers
21–22	3	22–24	31	21–25	54
23–24	28	25–27	75	26–30	108
25–26	56	28–30	56	31–35	108
27–28	42	31–33	77	36–40	35
29–30	33	34–36	43	41–45	19
31–32	50	37–39	18	46–50	20
33–34	47	40–42	11	51–55	4
35–36	23	43–45	13	Total	348
37–38	15	46–48	18		
39–40	8	49–51	4		
41–42	6	52–54	2		
43–44	9	Total	348		
45–46	15				
47–48	7				
49–50	2				
51–52	4				
Total	348				

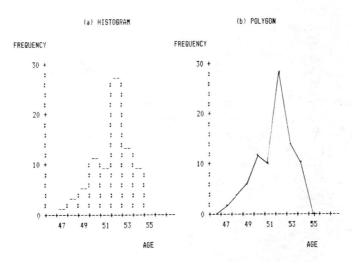

Figure 2-1. Histogram and Polygon for Senior Officer Ages.

Table 2-7. Distribution of Maximum Prices for Moorings at Shilshole Bay.

Price	Number of responses	Price	Number of responses	
			Equal or less	Greater
$2.01– 3.00	43	$ 2.00	0	755
3.01– 4.00	158	3.00	43	712
4.01– 5.00	239	4.00	201	554
5.01– 6.00	175	5.00	440	315
6.01– 7.00	60	6.00	615	140
7.01– 8.00	24	7.00	675	80
8.01– 9.00	14	8.00	699	56
9.01–10.00	6	9.00	713	42
over 10.00	36	10.00	719	36
Total	755		755	755

vertical bars correspond to the frequency of each value or class of values. The concentration of values at 52 years and the tailing off of values below 52 years are evident.

The frequency polygon, also shown in Figure 2-1, conveys the same information, but in a little more general fashion. Points are plotted above each value (or mid-values of classes) corresponding to the frequency of those values. Then they are connected by straight lines, and the figure is enclosed by extending the polygon line to the horizontal axis.

Cumulative distribution plots are often effective. Table 2-7 presents a distribution of prices (per foot per month) at which prospective clients for boat

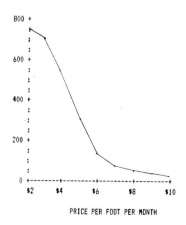

PRICE PER FOOT PER MONTH

Figure 2-2. Number of Customers Remaining at Increasing Prices.

spaces would remove their names from the waiting list at Shilshole Bay Marina near Seattle. [2]

In addition to counts in each class we show the number of responses that fall at or below selected prices and then the number that are greater than the same prices. Both these cumulations are obtained by adding (or cumulating) the frequencies from the original distribution. We show both the "equal or less than" and the "more than" cumulations here for completeness. Usually one would select one or the other to emphasize. For the mooring prices the numbers that would pay more than the listed amounts reflect the demand for moorings at different prices, so we show that graph in Figure 2-2. We see, for example, that at $5 demand for mooring space is less than half what it is at $2.

References

1. C. J. Verhoeven, *Techniques in Corporate Manpower Planning*, Nijhoff, Amsterdam, 1982, p. 50.
2. Courtesy of Susan Doolittle, Port of Seattle.

CHAPTER 3

Special Averages

One reason the mean is such a widely used average is that it lends itself to calculations of the following kind.

> Mr. Fix-it is repairing a boardwalk with $2'' \times 6''$ by 8-foot lumber. He initially bought 80 feet at 20 cents a running foot. To complete the project he later bought 40 feet at 25 cents a foot and 24 feet at 30 cents a foot. What was his cost per foot for the entire project?

Here, Fix-it's three purchase prices were 20, 25, and 30 cents, but his average price is not $(20 + 25 + 30)/3 = 25$ cents, because he bought different quantities at the several prices. The average purchase price per foot is easily found by calculating the total dollars paid and dividing by the total feet obtained.

$$
\begin{aligned}
20 \text{ cents per foot} \times 80 \text{ feet} &= \$16.00, \\
25 \text{ cents per foot} \times 40 \text{ feet} &= \$10.00, \\
30 \text{ cents per foot} \times 24 \text{ feet} &= \$\ 7.20, \\
144 \text{ feet} &= \$33.20, \\
\text{Average} = \$33.20/144 &= 23\tfrac{3}{4} \text{ cents.}
\end{aligned}
$$

The average purchase price of $23\frac{3}{4}$ cents can be viewed as a *weighted* average of the three purchase prices. This average is closer to 20 cents than 30 cents because more lumber was bought at the lower than at the higher prices.

Some of the averages that we will look at in this section are weighted averages, and the key to understanding and using them is to understand how weights work and what they can do.

Index Numbers

In 1972 the ingredients for a pound cake cost 87 cents a pound for butter, 14 cents a pound for sugar, 52 cents a dozen for eggs, and $1.20 for a 10-pound bag of flour. To make the pound cake requires 1 pound of butter, 1 pound of sugar, nine eggs, and 2 pounds of flour (add a teaspoon of vanilla and half a teaspoon of salt, whose costs we ignore). So the cost of making the pound cake in 1972 was

$$
\begin{array}{ll}
\text{1 lb of butter @ \$0.87 per lb} & = \$0.87. \\
\text{1 lb of sugar @ \$0.14 per lb} & = \$0.14, \\
\text{9 eggs at \$0.52 per dozen} & = \$0.39, \\
\text{2 lb of flour @ \$1.20 per 10 lb} & = \$0.24, \\
\text{Total cost} & = \$1.64.
\end{array}
$$

By 1983 the price of butter had risen 115 percent to $1.87 per lb, sugar had risen 93 percent to $0.27 per lb, eggs had gone up 69 percent to $0.88 per dozen, and flour had increased 129 percent to $2.75 per 10-lb bag. The cost of making the pound cake in 1983 was $1 \times \$1.87 + 1 \times \$0.27 + (9/12) \times \$0.88 + (2/10) \times \$2.75 = \$3.35$. The 1983 cost, $3.35, is 204 percent of the 1977 cost of $1.64. Records of average unit prices in the United States back to 1920 allow us to find historical costs of filling the pound-cake recipe. These are shown in Figure 3-1.

The price index for the cost of the cake is a *weighted* average of the price relatives for the four ingredients. The price relatives for 1983 compared to

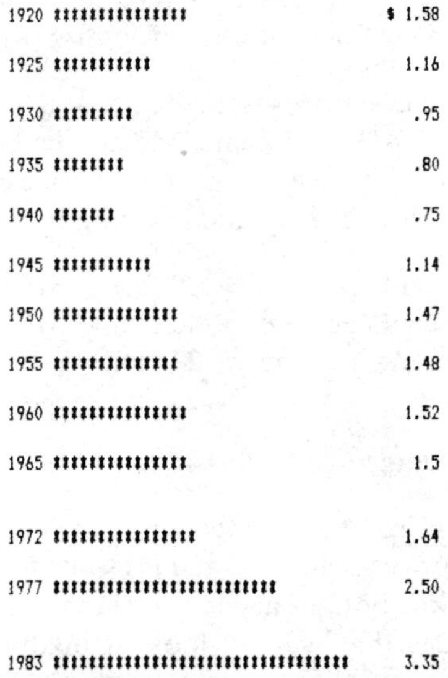

```
1920 ::::::::::::::::           $ 1.58

1925 ::::::::::::              1.16

1930 :::::::::                 .95

1935 ::::::::                  .80

1940 :::::::                   .75

1945 :::::::::::::             1.14

1950 :::::::::::::::::         1.47

1955 :::::::::::::::::         1.48

1960 :::::::::::::::::::       1.52

1965 ::::::::::::::::::        1.5

1972 ::::::::::::::::::::      1.64

1977 :::::::::::::::::::::::::::::::    2.50

1983 :::::::::::::::::::::::::::::::::::::::::::    3.35
```

Figure 3-1. Cost of Making a Pound Cake.

1972 for the individual ingredients are

$$\begin{array}{llll}
\text{Butter} & \$1.87/\$0.87 = 2.149, & \text{or } 214.9\%, \\
\text{Sugar} & \$0.27/\$0.14 = 1.929, & \text{or } 192.9\%, \\
\text{Eggs} & \$0.88/\$0.52 = 1.692, & \text{or } 169.2\%, \\
\text{Flour} & \$2.75/\$1.20 = 2.292, & \text{or } 229.2\%.
\end{array}$$

To average these comparative prices we must take account of the relative importance of the ingredients to the total cost of the cake in the base year 1972. These figures are

$$\begin{array}{llll}
\text{Butter} & \$0.87/\$1.64 = 0.5305, & \text{or } 53.05\%, \\
\text{Sugar} & \$0.14/\$1.64 = 0.0854, & \text{or }\ \ 8.54\%, \\
\text{Eggs} & \$0.39/\$1.64 = 0.2378, & \text{or } 23.78\%, \\
\text{Flour} & \$0.24/\$1.64 = 0.1463, & \text{or } 14.63\%, \\
\text{Total} & & 100.00\%.
\end{array}$$

Butter accounted for a little over one-half of the cost of making the cake in 1972, and will get a corresponding weight in figuring the price index. The final price index is a weighted average of the price relatives for the ingredients, where the weights are the relative importance figures that we have just calculated.

Ingredient	Price relative		Relative importance		Product
Butter	2.149	×	53.05	=	114.03
Sugar	1.929	×	8.54	=	16.47
Eggs	1.692	×	23.78	=	40.24
Flour	2.292	×	14.63	=	33.53
Total			100.00		204.27

We calculated the 204 in a more straightforward way to begin with. The reason for converting to price relatives and relative importance figures for the ingredients is to introduce the method of calculation of most of the world's important price indexes, including the Consumer Price Index of the Bureau of Labor Statistics (CPI of the BLS) in the United States.

The CPI was initiated during World War I when the Shipbuilding Labor Adjustment Board requested BLS to develop a fair way of adjusting wage rates in shipyards for changes in the cost-of-living. Prices were sampled in retail establishments in 32 cities and details on expenditures made by 12,000 wage-earner families in 92 cities were collected. Regular publication of a national consumer price index began in 1921. [1].

The CPI-W (price index for urban wage and clerical worker families and single individuals living alone) is the successor index to the one begun during World War I. The BLS also publishes CPI-U, which is an all urban residents (including salaried workers, self-employed, retirees, unemployed) price index.

These consumer price indexes are based on price relatives for 382 items of goods and services grouped into 68 detailed expenditure classes. From comparative price data collected in 2300 food stores and other points of purchase, price relatives are constructed for the detailed expenditure classes. Then these are weighted by relative importance measures derived from a base survey of consumer expenditure patterns. Except for the fact that the items priced to construct price relatives are only a sample of the items in an expenditure class, the calculation method is the same as the one shown above for the pound cake. But this is an important difference, because it means that the price index can only be derived by weighting price relatives by relative importance measures. There is not an alternative direct calculation method.

The interpretation of the CPI parallels the pound cake example, however. There we had the relative price change of the bill of goods, or market basket, needed to make the cake. In the CPI we have the relative price change of the market basket of goods and services consumed by the average urban resident (wage and clerical or all urban residents, depending on the index).

The relative importance of broad categories of goods and services in the CPI-U in 1982 is given in Table 3-1. Within the housing category, a detailed item of house furnishings and operation that would be subject to pricing would be a television set. Television sets have a relative importance of 0.256 percent. Flour and prepared flour mixes are specific items within the food category with a weight of 0.093 percent. The methods of almost all Western nations follow those of the United States.

Table 3-2 illustrates the calculation of a price index by the weighted average of relatives method. A local newspaper had collected the prices of food items in particular retail stores in late 1977. The specific items priced can be regarded as a sampling of items within particular broad classes of food, such as cereals and bakery products, dairy products, etc. The items priced in one store were

Table 3-1. CPI-U Market Basket by Major Expenditure Groups, 1982*

Expenditure group	% Total expenditures
Food and beverages	17.4
Housing	46.0
Apparel	4.5
Transportation	18.9
Medical care	5.2
Entertainment	3.7
Personal (including educational)	3.1
Other	1.2
Total	100.0

*SOURCE: BLS Bulletin 2183, September, 1983.

priced again in late 1984. The table shows the calculation of a food price index for customers of this store.

First, the relative for 1984 compared to 1977 price is shown for each food item. The relatives in each group are averaged to obtain a group price relative. Then the group price relatives are averaged, in turn, by weighting each group relative by the relative importance figure for the group. Relative importance is the percentage of the CPI-U market basket dollar expenditures devoted to the category in 1977. This is the calculation method of the BLS Consumer's Price Index. In the BLS indexes the groups of items are more detailed than shown in our example.

Another kind of price index compares places rather than times. The *Financial Times* of London published the results of a survey of "living costs" for businessmen traveling in different cities. The index was based on the cost of three nights bed and breakfast, two *à la carte* dinners in a first class international hotel, one dinner in an average restaurant, three bottles of house wine, one hotel lunch, two snack meals, five whiskeys, and a 5-kilometer taxi journey (to return to the hotel after five whiskeys we can assume!). Taking London as the base (100), the index was 96 for Paris, 76 for New York, 50 for Rome, and 42 for Johannesburg. [2]

Some History

William Stanley Jevons (1835–1882) is known mainly as one of the first neoclassical economists to emphasize utility as a foundation of economic theory. Jevons also pioneered a number of techniques in economic statistics. Among these were the analysis of economic time series into trend, cyclical, and seasonal components and the measurement of price changes by means of index numbers. In 1863 Jevons calculated indexes for English prices back to 1782. Jevons was an innovator in the use of graphs in economic statistics. J. M. Keynes observed that

> Jevons was the first theoretical economist to survey his material with the prying eyes and fertile, controlled imagination of the natural scientist. He would spend hours arranging his charts, plotting them, sifting them, tinting them nearly with delicate pale colours like the slides of the anatomist, all the time pouring over them to discover their secret. [3]

In 1864 Etienne Laspeyres (1834–1913), who had worked on price indexes for Hamburg, Germany, proposed the use of weighted aggregative price indexes. These are indexes using quantity weights in the manner of our pound cake example. Laspeyres proposed the use of fixed base year quantity weights. This remains the most popular index number method today. In 1874 Hermann Paasche (1851–1925) proposed the use of current year quantity weights in aggregative price indexes. The index formulas recommended by Laspeyers and Paasche are still often referred to as the Laspeyres and the Paasche indexes.

Table 3-2. Prices of Food at Home, 1977–1984.
Budget Barometer Price Index for Food Prepared at Home.

	Relative importance	Price 1977	Price 1984	Price relative	Group relative	Index calculation
Cereals and Bakery Products	12.50				1.61	20.12
Wheaties (18 oz)		$0.95	$1.89	$1.99		
Nabisco Oreo cookies (15 oz)		0.95	1.87	1.97		
Betty Crocker white cake mix (18.5 oz)		0.69	0.80	1.16		
Gold Metal flour (5 lb)		0.98	1.19	1.21		
American Beauty wide egg noodles (12 oz)		0.52	0.82	1.58		
Minute rice (14 oz)		0.76	1.33	1.75		
Meats, Poultry, Fish, and Eggs	32.23				1.84	59.23
Chicken, whole fryers (per lb)		0.45	0.83	1.84		
Beef round steak (bone-in) USDA Choice (per lb)		1.18	2.85	2.42		
Chicken of the Sea chunk light tuna (6.5 oz)		0.71	0.89	1.25		
Dairy Products	13.52				1.63	22.01
Kraft Velveeta pasteurized processed cheese spread (2 lb)		2.09	3.49	1.67		
Thatcher's Guernsey milk ($\frac{1}{2}$ gal)		1.04	1.65	1.59		

	Weight				
Fruits and Vegetables	14.38		1.91		27.40
Yellow onions (per lb)		0.13	0.25	1.92	
Iceberg lettuce (per lb)		0.33	0.79	2.39	
Green Giant Niblets, whole kernel corn (12 oz)		0.33	0.55	1.67	
Mountain Pass chopped green chilli (4 oz)		0.36	0.54	1.50	
Betty Crocker mashed potato buds (16.5 oz)		0.99	1.50	1.52	
Del Monte fruit cocktail (17 oz)		0.47	1.20	2.55	
Minute Maid frozen orange juice (6 oz)		0.33	0.59	1.79	
Other Foods at Home	27.37		1.57		43.07
Blue Bonnet margarine (1 lb)		0.56	0.79	1.41	
Del Monte ketchup (20 oz)		0.59	1.04	1.76	
Nestlé's semi-sweet chocolate chips (12 oz)		1.16	2.29	1.97	
ReaLemon reconstituted lemon juice (24 oz)		0.83	1.58	1.90	
Hamburger Helper (6.5 oz)		0.69	0.94	1.36	
Wesson Oil (24 oz)		0.92	1.77	1.92	
Nestea instant tea mix (3 oz)		1.69	2.39	1.41	
Peter Pan peanut butter, smooth (12 oz)		0.72	1.25	1.74	
Alpo dog food beef chunks (14.5 oz)		0.35	0.42	1.20	
Hills Bros coffee, regular grind (1 lb)		2.89	2.79	0.97	
Cane sugar (5 lb)		1.04	1.99	1.91	
Hunts tomato sauce (8 oz)		0.20	0.25	1.25	
Campbell's cream of mushroom soup (10.5 oz)		0.25	0.41	1.64	
Total	100.00				

Index = 171.84

Francis Y. Edgeworth (1845–1926), in the period 1887–1890, conducted a study of prices indexes for the British Association for the Advancement of Science. He used a variety of price index formulas. In 1893 Roland Faulkner in a study commissioned by the U.S. Senate constructed price indexes for the United States from 1841 to 1891. In all of the studies mentioned the prices involved were wholesale or producer prices. In 1902 the Bureau of Labor Statistics began its index of wholesale prices in the United States.

In 1922 Irving Fisher (1867–1947), a respected economist, published his study *The Making of Index Numbers.* [4] He compiled 134 formulas that had been used or suggested for constructing price indexes. These involve the use of different construction methods such as averages of relatives versus relatives of aggregates, different methods of averaging such as mean versus median versus other averages, and different methods of weighting such as Laspeyres versus Paasche. He devised various tests of consistency for index number formulas— such as that they give the same relation between prices at different periods by working forward as by working backward (base-reversal test). Fisher concluded that the square root of the product of the weighted aggregate price index of Laspeyres (base year weights) and Paasche (current year weights) was the best index number formula. While Fisher cites some earlier advocates of this index, it has since been known as *Fisher's ideal index.* Almost all the widely used indexes in the world, however, use fixed weights of a historic, though not necessarily base year, period.

The Harmonic Mean

An average with some special properties and uses is the harmonic mean. Here is an example:

> The Road Warrior has embarked on a 100-mile trip. For the first 50 miles the warrior travels at 50 mph, but is slowed to 25 mph for the second 50 miles owing to road repairs. Does the Road Warrior average 37.5 mph?

A little thought reveals that the average is not 37.5 mph. The Road Warrior took 1 hour to cover the first 50 miles and 2 hours to travel the second 50 miles. The average speed must be 100 miles divided by 3 hours, or 33 mph. The weighted average of speeds, where the weights are the hours at the different speeds is one way to calculate.

$$50 \text{ mph for 1 hour } = 50 \text{ miles,}$$
$$25 \text{ mph for 2 hours } = 50 \text{ miles,}$$
$$\text{Total for 3 hours } = 100 \text{ miles,}$$
$$\text{Average speed } = 100/3 = 33 \text{ mph.}$$

The harmonic mean of X is the reciprocal of the average of the reciprocals of X. For 50 and 25,

$$1/50 = 0.02,$$
$$1/25 = 0.04,$$
$$\text{Total} = 0.06,$$
$$\text{Average of the reciprocals} = 0.06/2 = 0.03,$$
$$\text{Reciprocal of } 0.03 = 1/0.03 = 33.$$

The reason that the harmonic mean is the correct average here is that the *numerators* of the original ratios to be averaged were equal.

$$50 \text{ miles in 1 hour } = 50 \text{ mph},$$
$$50 \text{ miles in 2 hours } = 25 \text{ mph}.$$

When the *denominators* of ratios to be averaged are equal we can use the ordinary (or arithmetic) mean to get the average. For example,

$$\text{Travel an hour at 50 mph},$$
$$\text{Travel an hour at 25 mph},$$
$$\text{Average speed} = 37.5 \text{ mph},$$

When the denominators are not equal you can use the weighted mean. That is what we did at first with the average speed for traveling 50 miles at 50 mph and 50 miles at 25 mph. That is also the special case of equal numerators for which the harmonic mean works.

Averaging Time Rates

When rates are connected in a time sequence, special methods of averaging are called for. Here is an example:

A new business has a satisfactory sales volume in its first year. In the second year sales increase by 100 percent (over the first year). Then in the third year sales decrease 75 percent (from the second year level). Is it true that sales are still 25 percent above the first year's level?

The best way to answer a question like this is to figure it out. Suppose sales were $100,000 in the first year. A 100 percent increase takes them to $200,000 the second year. Then, a 75 percent decrease take them *down* $150,000 to $50,000. Sales volume in the third year is actually only 50 percent of first year volume.

The secret is that successive time relatives combine through multiplication rather than addition. We have

$$\frac{X(2)}{X(1)} \times \frac{X(3)}{X(2)} = \frac{X(3)}{X(1)},$$

or

$$2.0 \times 0.25 = 0.50.$$

To combine changes in situations like this we have to multiply time relatives. To average changes we have to take the appropriate root of the relative for the entire time period. An example of combining changes is compound interest in investments. If $2000 is invested for 5 years at 20 percent interest, the investment (compounded) grows to

$$\$2000(1.20)^5 = \$2000 \times 2.488 = \$4976.$$

If you invested $2000 for 5 years and wound up with $3600, your average yearly multiple would be

$$(\$3600/\$2000)^{1/5} = (1.8)^{0.2} = 1.12$$

and the average yearly return is $(1.12 - 1.00) \times 100 = 12\%$.

Let us look at some averages from the series on the cost of making the pound cake. In 1935 the cost was $0.80 and in 1965 the pound cake cost $1.50 to make. What was the average increase in price per 5-year period?

$$\text{Average multiple} = (\$1.50/\$0.80)^{1/6} = 1.875^{1/6} = 1.1105,$$
$$\text{Average rate} = (1.1105 - 1.000) \times 100 = 11.05\%.$$

How about the average price increase per year from 1935 to 1983? The answer is

$$\text{Average multiple} = (\$3.35/\$0.80)^{1/48} = 4.1875^{1/48}$$
$$= 1.0303,$$
$$\text{Average rate} = (1.0303 - 1.0000) \times 100 = 3.03\%.$$

Simpson's Paradox

The change in a weighted average can be as much the result of changes in the weights as it is of changes in the values being averaged. An extreme case of what can happen is called Simpson's paradox, apparently after E. H. Simpson, who called attention to the possibility in a 1951 article. [5] The paradox is illustrated here by a real example from income taxes. [6]

From 1974 to 1978 the effective income tax rates in each of five broad income tax brackets went down, but yet the overall effective income tax rate increased. The effective income tax in an income bracket is the total income tax divided by the total income earned by recipients in the bracket. The figures are given in Table 3-3.

We see that over the period the percentage of income earned by persons in the brackets under $15,000 decreased, while the share of total income earned by persons in the over $15,000 brackets increased. With the higher brackets

Table 3-3. Income Tax Rate and Percent of Income Earned by Income Category, 1974 and 1978.

Adjusted gross income	Tax rate		% Total income	
	1974	1978	1974	1978
Under $ 5000	0.054	0.035	4.73	1.60
$ 5000–$ 9999	0.093	0.072	16.63	9.89
$10000–$14999	0.111	0.100	21.89	13.83
$15000–$99999	0.160	0.159	53.40	69.62
$100000 & more	0.384	0.383	3.34	5.06
Total	0.141	0.152	100.00	100.00

having higher tax rates, the overall (or weighted average) rate tends to increase even in the absence of change in the effective tax rates within brackets. Over the period in question the effective rates declined bracket by bracket, but the effect of the change in weights was enough to produce an increased overall tax rate notwithstanding. Inflation over the period was causing "bracket creep."

References

1. "The Consumer Price Index: Concepts and Content Over the Years," BLS Report 517, May 1978 (Revised).
2. *The Week in Review*, Deloitte, Haskins, & Sells, March 7, 1980.
3. Paul J. Fitzpatrick, "Leading British Statisticians of the Nineteenth Century," *Journal of the American Statistical Association*, **55**, No. 289 (March 1960), 55.
4. Irving Fisher, *The Making of Index Numbers*, Houghton Mifflin, Boston, 1923, p. 459.
5. E. H. Simpson, "The Interpretation of Interaction in Contingency Tables," *Journal of the Royal Statistical Society*, Ser. B, **13** (1951), 238–241.
6. Clifford R. Wagner, "Simpson's Paradox in Real Life," *American Statistician*, **36**, No. 1 (February 1982), 48.

CHAPTER 4

Making Comparisons

In 1954 Branch Rickey wrote an article for *Life Magazine* entitled "Goodbye to Some Old Baseball Ideas." [1] Rickey criticized some traditional baseball statistics and proposed some of his own, which included formulas for team offense and pitching efficiency. For individual hitting Rickey recommended the sum of *on-base average* (OBA) plus *extra base power* (EBP). These were defined as

$$\text{OBA} = \frac{\text{hits} + \text{bases on balls} + \text{hit by pitcher}}{\text{at bats} + \text{bases on balls} + \text{hit by pitcher}},$$

$$\text{EBP} = \frac{\text{total bases} - \text{hits}}{\text{at bats}}.$$

Rickey explained that OBA was better than the batting average because the batting average excluded bases on balls (from both numerator and denominator) for example. One objective of a hitter is to get on base, and great hitters have always received a large number of bases on balls, either intentionally or because the pitcher is extremely careful. He preferred EBP to the more traditional slugging average (total bases/at bats) because EBP gets more at the essence of power by subtracting the hits. It is the number of extra bases per time at bat. For Rickey, hits are already included in the on-base average.

Comparing Observations

Rickey presented his list of greatest hitters since 1920. It is reproduced in Table 4-1 with the OBA, EBP, and the batting rating (OBA + EBP). Babe Ruth ranks first, followed by Ted Williams, Lou Gehrig, and Jimmy Foxx.

Table 4-1. Branch Rickey's Greatest Hitters.

Hitter		Points out of 1000			Deviations from mean		
		OBA	EBP	SUM	OBA	EBP	SUM
1	Babe Ruth	481	271	752	65.48	94.04	159.52
2	Ted Williams	484	218	702	68.48	41.04	109.52
3	Lou Gehrig	447	219	666	31.48	42.04	73.52
4	Jimmy Foxx	429	213	642	13.48	36.04	49.52
5	Rogers Hornsby	449	185	634	33.48	8.04	41.52
6	Hank Greenberg	412	219	631	−3.52	42.04	38.52
7	Ralph Kiner	404	212	616	−11.52	35.04	23.52
8	Stan Musial	432	177	609	16.48	0.04	16.52
9	Joe Dimaggio	398	191	589	−17.52	14.04	−3.48
10	Mel Ott	414	172	586	−1.52	−4.96	−6.48
11	Charlie Keller	410	174	584	−5.52	−2.96	−8.48
12	Johnny Mize	397	187	584	−18.52	10.04	−8.48
13	Harry Heilman	431	148	579	15.48	−28.96	−13.48
14	Tris Speaker	441	135	576	25.48	−41.96	−16.48
15	Hack Wilson	395	178	573	−20.52	1.04	−19.48
16	Ken Williams	400	168	568	−15.52	−8.96	−24.48
17	Earl Averill	395	162	557	−20.52	−14.96	−35.48
18	Roy Campanella	375	178	553	−40.52	1.04	−39.48
19	Lefty O'Doul	413	139	552	−2.52	−37.96	−40.48
20	Bob Johnson	393	157	550	−22.52	−19.96	−42.48
21	Chuck Klein	379	167	546	−36.32	−9.96	−46.48
22	Dolph Camilli	382	161	543	−33.52	−15.96	−49.48
23	Ty Cobb	431	111	542	15.48	−65.96	−50.48
24	Jackie Robinson	414	125	539	−1.52	−51.96	−53.48
25	Tommy Henrich	382	157	539	−33.52	−19.96	−53.58
	Mean	416	177	592			
	Standard deviation	28.5	34.4				

Some might ask why Rickey adds the on-base averages, which average around 400 (thousandths) and extra base power ratios, which are figures around 200, to obtain the batting rating. Does that give twice as much weight to OBA as to EBP? The answer is that it is not the magnitude of the measures, but their *variation* that gives the measures weight in distinguishing among the hitters. If all the hitters had about the same extra base power (EBP), then that element of hitting would not affect the variation in the batting rating. Variation in the batting rating would then follow the variation in the on-base average—the only respect in which the hitters differed.

A measure of variation is the standard deviation, and we calculated it for each of the batting variables. The standard deviations are 28.5 for the OBAs and 34.4 for the EBPs, so there is not a great difference in variation.

The next question to ask is whether the elements are really any different.

After all, if batters who were high in on-base average were also high in extra base power, then we would not need both measures to rate the batters. Either one would do as well. To focus on this question we have expressed the two variables as deviations from their means in Table 2-1. We see that while Ruth excelled in both OBA and EBP, Hornsby earns his high ranking from a high OBA and near average EBP, and Greenberg earns his with a high EBP and near average OBA. From these examples and others we can see that the two measures do not go together. OBA and EBP are not like Tweedledum and Tweedledee.

From this example we see that it is variation that counts. To say that Ted Williams has a hitting rating 109 points above the average for the 25 men which is made up of 68 points above the mean on base average and 41 points above the mean extra base power tells us more than the statement that his batting rating is 484 + 218 = 702. Scores related to the averages mean more than scores in isolation.

Standard Scores

In another example we carry this principle of comparison a step further. Table 4-2 contains data on the distribution of employment by economic sectors for 21 OECD countries. [2]

How can we effectively characterize the sector profile for each country? We might calculate means and then express each country's share of employment in the sectors as deviations from mean shares, just as we did for the batting measures. But here the standard deviations differ considerably. The standard deviations are 18.0 for agriculture shares, 11.4 for industry, and 9.5 for service.

In the batting rating comparisons we calculated deviations from means. This eliminated the different means by subtraction. If we now average the deviations from the mean, we will get zero for each series. The subtraction transformed each series into one with a mean of zero. We did not do anything about the standard deviations because they were not that different. In the sector employment data we desire to adjust for the different standard deviations as well. The adjustment, or transformation, is

$$Z = \frac{X - \bar{X}}{S}.$$

We are going to divide each deviation by the standard deviation. The result is the deviation of each value from the mean, expressed in standard deviation units. These values are shown in the second three columns for the 21 nation employment data.

The Z-transformation creates a variable with a mean of zero and a standard deviation of 1.0. The set of Z-scores for the employment sector data give us a quick profile for each country. Generally, Z-scores toward ± 2.0 place an

Table 4-2. Sector Distribution of Employment in 21 OECD Countries, 1960.

Country	% Total employment			Z-Scores for			Ranking for		
	AGR	IND	SER	AGR	IND	SER	AGR	IND	SER
United States	8	38	54	−0.93	−0.46	2.30	19	15	1
Canada	13	43	45	−0.66	−0.02	1.35	16	14	2
Sweden	14	53	33	−0.60	0.86	0.08	15	4	10.5
Switzerland	11	56	33	−0.77	1.13	0.08	17.5	2.5	10.5
Luxembourg	15	51	34	−0.54	0.69	0.19	13.5	6	8.5
United Kingdom	4	56	40	−1.16	1.13	0.82	21	2.5	4.5
Denmark	18	45	37	−0.38	0.16	0.50	12	12	6
West Germany	15	60	25	−0.54	1.48	−0.76	13.5	1	17
France	20	44	36	−0.27	0.07	0.40	10.5	13	7
Belgium	6	52	42	−1.05	0.78	1.03	20	5	3
Norway	20	49	32	−0.27	0.51	−0.03	10.5	7.5	12.5
Iceland	25	47	29	0.01	0.34	−0.34	8	9.5	15
Netherlands	11	49	40	−0.77	0.51	0.82	17.5	7.5	4.5
Austria	23	47	30	−0.10	0.34	−0.24	9	9.5	14
Ireland	36	30	34	0.63	−1.16	0.19	5	19	8.5
Italy	27	46	28	0.12	0.25	−0.45	7	11	16
Japan	33	35	32	0.46	−0.72	−0.03	6	17	12.5
Greece	56	24	20	1.74	−1.69	−1.29	2	20	20
Spain	42	37	21	0.96	−0.55	−1.19	4	16	19
Portugal	44	33	23	1.07	−0.90	−0.98	3	18	18
Turkey	79	12	9	3.02	−2.75	−2.45	1	21	21
Mean	24.8	43.2	32.2	0.0	0.0	0.0			
Standard deviation	18.0	11.4	9.47	1.0	1.0	1.0			

observation toward the extreme of the distribution for the variable, while those within one standard deviation of the mean are more or less typical of the group.

For example, we see that Turkey's agricultural share of employment is 3.0 standard deviations above the mean share for the 21 countries, while its shares of industry and service employment are 2.75 and 2.45 standard deviations below average. Iceland is very average, with Z-scores of 0.01, 0.34, and −0.34 for the three sectors. The Z-scores for the United States of −0.93, −0.46, and 2.30 tell us that the United States is quite low in agricultural share, moderately low is industrial share, but very high in service share.

Comparing Variables

In the two examples so far we have been comparing the hitters and the countries. That is, we have been comparing the *subjects*, or observational units in the studies. We found that we could make effective descriptive comparisons of the subjects by adjusting for certain differences in the *variables*.

Table 4-3. Distributions of Sector Shares, 21 OECD Countries, 1960.

Percent of employment	Number of countries		
	Agriculture	Industry	Service
0– 9	3	—	1
10–19	7	1	—
20–29	5	1	6
30–39	2	5	9
40–49	2	8	4
50–59	1	5	1
60–69	—	1	—
70–79	1	—	—
Total	21	21	21
Mean	24.8	43.2	32.2
Standard deviation	18.0	11.3	9.5

We often look for effective means of summarizing the variables themselves. In such cases we want to emphasize the differences. The methods that we have already seen are the frequency distribution (table or graph) and the mean and standard deviation. These are shown in Table 4-3 for the sector share distributions. We see that service sector employment shares are least variable and agricultural employment shares are most variable. A further characteristic is called *skewness*. We have referred to it before, but not by this technical term. For example, the pattern of the agriculture share distribution is that the most typical value is low in the range (10–19) with frequencies stretching out toward the higher values. This is termed positive skewness. Galton's honors examination scores had very high positive skewness, you might recall.

Median and Quartiles

In Table 4-2 we gave the rankings of the countries on each employment share variable. The ranks are another way to summarize the standings for a country. Thus, the United States is nineteenth out of 21 countries in agriculture share, fifteenth in industry share, and first in service share of employment. Ranks serve to define another group of measures for frequency distributions. These measures are simply particular positional locations. Out of 21 values the middle one in order of magnitude is the eleventh. This middle value is called the *median*. One can also locate a value one-quarter of the way through the ranking and a value three-quarters of the way through the ranking. When a list of values is ordered from low to high, these are called the *first* and the *third quartile.* The three positional measures are defined as the values of the observations with the following ranks.

	Ranking	
	Low to high	High to low
First quartile	$(n + 1)/4$	$3(n + 1)/4$
Median	$(n + 1)/2$	$(n + 1)/2$
Third quartile	$3(n + 1)/4$	$(n + 1)/4$

With 21 values ranked from high to low, the third quartile is designated as the value having the $5\frac{1}{2}$ rank, the median is the value having the eleventh rank, and the first quartile is the value having the $16\frac{1}{2}$ rank. When a fractional rank is indicated we take the mid-value of the adjacent observations.

These positional measures, along with the mean, are given below for each employment share variable.

	Agriculture	Industry	Service
First quartile	12	36	26.5
Median	20	46	33
Third quartile	34.5	51.5	38.5
Mean	24.8	43.2	32.2
Median	20.0	46	33

Certain comparisons among the measures listed above can be used as indicators of skewness in distributions. The first is the distance from the first quartile to the median compared with the distance from the median to the third quartile. If the pattern of the distribution is positive skewness, the distance from the median to the third quartile will exceed that from the first quartile to the median. The highest densities occur early. When skewness is negative the higher frequency densities are in the upper ranges and the frequencies stretch out toward the lower values. This causes the distance from the first quartile to the median to exceed the distance from the median to the third quartile.

From these comparisons we find that shares of employment in agriculture are positively skewed, while shares of employment in industry are negatively skewed. In service shares the skewness is less marked. Figure 4-1 shows the difference graphically.

The second indication of skewness is found in a comparison of the mean and the median. The mean is like a balance point and will be "pulled" in the direction of skewness by any stretching out of frequencies toward one extreme that is not balanced by an equal stretching out to the other extreme. Thus,

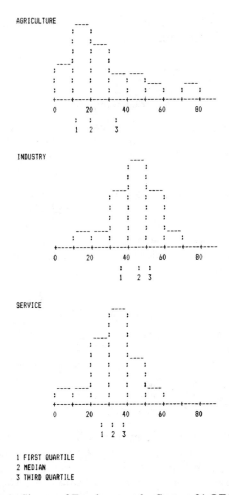

Figure 4-1. Percentage Shares of Employment by Sector, 21 OECD Countries, 1960.

when the mean exceeds the median there is positive skewness (agriculture) and when the mean is less than the median there is negative skewness (industry). The mean and median are quite close for the service shares.

Boxplots

A graphic device that presents the same material as histograms or polygons is the boxplot. The boxplot is based on the positional measures just described and the extremes of a set of values. Figure 4-2 shows boxplots for our sector

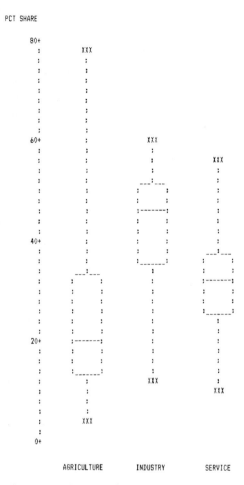

Figure 4-2. Boxplots for Sector Share Data.

share distributions. A vertical box extends from the first to the third quartile, with a line drawn across at the median. Then, a single vertical line is extended in either direction to the extremes of the distribution. If the median line is halfway through the box, the central portion of the distribution is essentially symmetrical (not skewed). By noting the lengths of the line segments extending from the box toward the extremes, we can note how the densities of the extreme quarters of the distribution compare. In many cases these will follow the middle half. In the agriculture plot both the box and the line segments (sometimes called whiskers) stretch out toward the high values. In the industry plot they both stretch out toward the low values, and in the service plot both aspects are more symmetrical.

Figure 4-3 shows a variety of frequency distributions that were presented in a text by L. H. C. Tippett. [3] The height distribution is of U.S. Army recruits (taken from Pearson, 1895), the leaves per whorl are for *ceratophyllum* (Pearl, 1907), letters per word are from the *Concise Oxford Dictionary*, the sizes of firms are for spinning firms in Britain from the 1930 Census of Production, degrees of cloudiness are for days in July from 1890–1904 in Greenwich, U.K. (Pearse, 1928), and the number of rays are the frequency of flowers in *Chrysanthemum leucanthemum* (Tower, 1902). With each of Tippett's frequency graphs we have drawn the equivalent boxplot.

A variety of shapes is shown. The short box and long whiskers for the heights reflects high central concentration. Leaves per whorl and letters per

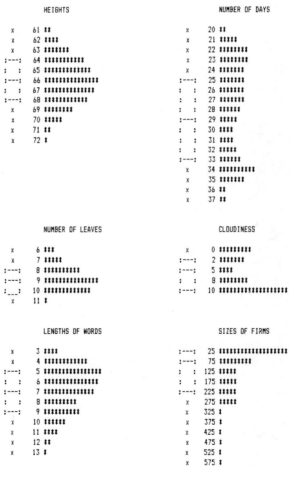

Figure 4-3. Frequency Distributions.

word have symmetry in the central half of the distributions with an excess tail to the left for the leaves and to the right for the words in the extreme half of the distribution. Sizes of firm shows high positive skewness in the extremes of the distribution. The fact that the length of the box exceeds the length of the whiskers reflects the pattern of high densities toward the extremes in the degrees of cloudiness. Days tend to be all cloudy or all clear, more often cloudy. The boxplot fails to reflect the double concentration in the numbers of flowers around 22 and 32 flowers.

Ordinal Data

CBS newsman Robert Pierpoint, aided by John Chancellor of NBC, Frank Reynolds of ABC, and others compiled a set of ratings of the press conference behavior of U.S. presidents. Presidents were rated from 1 to 10 on a number of attributes. [4] The results appear in Table 4-4. Numbers such as this are quite different from the measures that we have seen so far in our examples. The numbers here serve only to order or rank the presidents on the various attributes. From the ratings we know that Kennedy ranked first in perceived combative skill, Nixon and Reagan next, then Carter and Johnson, and Ford and Eisenhower last. We do not know if the combative skill of Nixon (8) exceeds Carter's (6) by the same margin that Carter (6) exceeds Ford (4). If we knew that equal differences in values represented equal amounts of a property in question, then the scale would have qualified as an *interval* measure of the property. Thus equal differences in values on the Celsius or the Farenheit temperature scales really represent equal differences in heat. Rarely do we know if this feature is true for subjective rating scales. When measures serve only to rank observations they are called *ordinal* measures or scales.

When scales serve only to rank observations some statisticians prefer to analyze the ranks rather than the scale values. In the presidential ratings they would establish rankings for each attribute, and then total the rankings if an

Table 4-4. Ratings of Presidents.

	Candor	Informative value	Combative skill	Humor	Total
Eisenhower	8	7	4	3	22
Kennedy	5	6	9	9	29
Johnson	2	3	6	4	15
Nixon	5	6	8	4	23
Ford	7	5	4	6	22
Carter	8	7	6	7	28
Reagan	6	4	8	8.5	26.5

overall rating is desired. From Table 4-4 the rankings are

	Candor	Informative value	Combative skill	Humor	Total
Eisenhower	1.5	1.5	6.5	7	16.5
Kennedy	5.5	3.5	1	1	11
Johnson	7	7	4.5	5.5	24
Nixon	5.5	3.5	2.5	5.5	17
Ford	3	5	6.5	4	18.5
Carter	1.5	1.5	4.5	3	10.5
Reagan	4	6	3.5	2	15.5

The overall ranking based on the total ranks, compared with the ranking based on the total of the ratings assigned in Table 4-4 is

	Rank based on	
	Average rank	Average rating
Carter	1	2
Kennedy	2	1
Reagan	3	3
Eisenhower	4	5.5
Nixon	5	4
Ford	6	5.5
Johnson	7	7

The difference in results is small. Carter and Kennedy switch places as do Eisenhower and Nixon. At least these switches are kept within political parties!

Rating presidents is a popular activity. Professor Robert K. Murray of Pennsylvania State University and a Ph.D. candidate in history, Tim H. Blessing, conducted a survey among historians who held Ph.D.s and professorial rank in American universities. [5] Presidents were ranked on a six-point scale from 1–"great" to 6–"failure." Ratings were averaged and then ranked. Lincoln, Franklin D. Roosevelt, and Washington headed the list, followed by Jefferson, Theodore Roosevelt, and Wilson. The standard deviations of the ratings were also given as an indication of controversiality of judgments. The most controversial were Nixon, Lyndon B. Johnson, and Hoover, with standard deviations exceeding 1.0 rating point. By contrast, there was greatest unanimity on Lincoln, Franklin D. Roosevelt, and Washington with standard deviations approaching 0.5 rating points.

Table 4-5. Adjusting Per Capita Disposable Income for Price Changes.

Year	Disposable income	CPI-W 1967 = 100	Adjusted disposable income
1972	$3837	125.3	$3837/1.253 = $3062
1976	$5477	170.5	$5477/1.705 = $3212
1980	$8012	247.0	$8012/2.470 = $3243

The Dollar Scale

Per capita disposable income (personal income less federal, state, and local taxes) in the United States in 1972, 1976, and 1980 was as follows:

1972	1976	1980
$3837	$5477	$8012

This is a healthy increase. But we know that inflation during the 1970s eroded the purchasing power of the dollar. So the increase in dollar income is not a good measure of increase in material well-being. Because of changes in what it will buy, the dollar is an unreliable scale for measuring quantity changes in the economy.

But we saw earlier how price changes are measured in the form of price indexes—a special kind of average. The Consumers Price Index (CPI-W) measures changes in the price of a fixed market basket of goods and services purchased by the typical moderate income urban wage and salary worker. We can use the CPI figures to adjust the dollar income figures for price changes. This is done in Table 4-5. The income figures have been adjusted by dividing each year's figure by the decimal equivalent of the CPI index for that year. The result, because the base of the CPI is 1967, is to restate the income figures in terms of dollars of 1967 purchasing power. The $5477 actual per capita income in 1976 will buy the goods and services that would have cost $3212 in 1967. Thus, the material level of living increased by 150 1967 dollars. The change in income adjusted for price changes is sometimes called the change in *real* income. Between 1976 and 1980 we see a very small change in real income.

References

1. Branch Rickey, "Goodbye to Some Old Baseball Ideas," *Life Magazine*, August 2, 1954.

2. From J. Singleman, *From Agriculture to Services*, Sage, Beverly Hills, 1978, p. 11.
3. L. H. C. Tippett, *The Methods of Statistics*, 4th edn, Wiley, 1952, pp. 22–24.
4. Lloyd Shearer, "Pierpoint's Presidential Report Card," *Parade Magazine*, January 7, 1982.
5. "Ranking of Presidents by American Historians," *The Washington Post National Weekly Edition*, February 6, 1984.

CHAPTER 5

Probability

Modern probability theory had its start in 1654 when a French nobleman known as Chevalier de Mere wrote a letter to Blaise Pascal (1623–1662), a celebrated mathematician. The letter concerned some problems de Mere had encountered at the gaming tables. Pascal initiated a correspondence with another mathematician, Pierre de Fermat (1601–1655) about these problems. Neither Pascal nor Fermat ever published any of his work on probability, but most of the correspondence has survived.

Three years later a Dutch physicist and mathematician, Christian Huygens (1629–1695) published a pamphlet entitled "On Reasoning in Games of Chance." This was the first printed work on the subject, and it served to stimulate James Bernoulli. [1]

James Bernoulli (1654–1705) was the eldest son of a merchant-banker, Nicolas Bernoulli, who had settled in Basel, Switzerland after leaving Antwerp for religious reasons. Just when Bernoulli took an interest in probability and games of chance is not known, but in 1885 he published a number of journal papers on the subject that were obviously inspired by Huygen's work. [2]

Bernoulli was probably working on his *Ars Conjectandi* (The Art of Conjecture) in the 1690s, but it was not published until 1705, 8 years after his death. It was edited by his nephew, Nicolas Bernoulli (1687–1759). In *Ars Conjectandi* Bernoulli includes comments and solutions on Huygen's work, treats permutations and combinations and their applications to games of chance, and finally includes some ideas about extensions to civil, economic, and moral affairs.

After Bernoulli's death, but before the publication of *Ars Conjectandi*, Pierre de Montmort (1678–1719) published a work *Essay on the Analysis of*

Games of Chance. Abraham De Moivre (1667–1754) published a work in 1711 (*de Mensura Sortis*) which was expanded into *The Doctrine of Chances: or A Method of Calculating the Probabilities of Events in Play*, published in 1718. Montmort's work appealed to mathematicians, but De Moivre's was essentially a gambler's manual. [3]

There does not appear to be any major new contribution to the theory of probability during the remainder of the eighteenth century. In the early nineteenth century, however, we encounter Pierre Simon, Marquis de Laplace (1749–1827), who is referred to by Arne Fisher as "that resplendent genius in the investigation of the mathematical theory of chance." [4]

Laplace was an eminent astronomer and mathematician. His two great works were *Celestial Mechanics* and *Analytical Theory of Probability*. Laplace wrote an introduction to his work on probability that employed no mathematical symbols and was addressed to the general reader. [5] Laplace presents ten *principles* of probability calculations as a general introduction before he goes on to discuss applications in "natural philosophy," "moral sciences," and particular areas such as vital statistics and insurance. This material, Laplace says,

> is the development of a *Lesson on Probabilities*, which I gave in 1795 at the Normal Schools, where I was called as Professor of Mathematics with Lagrange, by a decree of the National Convention. I was to present, without the aid of analysis (mathematics) the principles and general results of the theory of probabilities expounded in this work, in applying them to the most important questions of life, which are only in effect, for the most part, problems in probability. [6]

The laws, or theorems, of probability calculations have not changed since Laplace. So we will use his principles as our guide to the calculus of probabilities.

Laplace's Principles of Probability

Principles 1 and 2 concern the measure of probability. Having reduced all "events of the same kind" to equally possible cases, the ratio of the number of cases "favorable" to the event (whose probability is sought) to the number of all cases possible is the measure of probability. If the various "cases" are not equally possible, we have to determine their possibilities, and then the probability in question is the sum of the probabilities of the favorable cases. How to do this is given by other principles.

Examples given by Laplace, and by textbooks since, are that the probability of heads on a toss of a coin is $1/2$, or that the probability of an ace on a throw of a die is $1/6$.

Principle 3 says that: "if two events are independent of one another, the probability of their combined existence is the product of their respective probabilities," Thus, the probability of throwing two aces with two dice is $(1/6)$ $(1/6) = 1/36$. The probability of n repetitions of the same event is the probability of the simple event raised to the nth power. Laplace gives the example of 20 successive witnesses transmitting an account, where the probability that each transmits it correctly is 0.9. The probability that the account will be correct after 20 transmissions (not given by Laplace) is

$$0.9^{20} = 0.1216.$$

Laplace then comments that historical events seen across many generations and regarded as certain "would be at least doubtful if submitted to this test."

One question that Chevalier de Mere put to Pascal can be answered with these first few principles. It had been gambling lore at the time that if a gambling house bet even money that a player would throw at least one six in four throws of a die the house stood to make money. From Principle 3 we see that the probability of *not* throwing a six four times in a row is

$$(5/6)^4 = 625/1296 = 0.482,$$

so the probability of throwing *at least one* six is $1 - 0.482 = 0.518$. There is a small margin in favor of the house for an even money bet.

Chevalier de Mere had been betting that a player would throw *at least one 12* in 24 tosses of two dice, and had been losing money. He put the question to Pascal what the actual probability was. The probability of a 12 on a throw of a pair of dice is

$$(1/6)(1/6) = 1/36,$$

so the probability of *not* throwing a 12 is 35/36. The probability of *not throwing a 12 twenty-four successive times* is

$$(35/36)^{24} = 0.5086,$$

and the probability of throwing *at least one 12 in twenty-four throws* is $1 - 0.5086 = 0.4914$. This is less than 0.50, and a bet on this event at even money is a losing bet in the long run.

Principle 4 and 5 have to do with dependent combinations of events. Laplace gives essentially the following example.

The are three balls in an urn. Two are white and one is black. One of the three is to be drawn out. Then one of two remaining is to be drawn. What is the probability that the two white balls will be drawn?

The probability of drawing a white ball on the first draw is 2/3. The probability of drawing a white ball on the second draw, *given that a white ball was drawn on the first draw*, is 1/2 because under the condition italicized only one of the two balls remaining in the urn is white. The probability of the combined event of two white balls being drawn is

$$(2/3) * (1/2) = 1/3.$$

In general, "when two events depend on each other, the probability of the compound event is the product of the probability of the first event and the probability that, this event having occurred, the second will occur."

At this point let us introduce some symbols, namely P(A) for the probability of one event and P(B) for the probability of another event. Laplace's Principle 3 is

$$P(A \text{ and } B) = P(A) * P(B) \qquad \text{for independent events.}$$

Using the notation P(B|A) for the probability of event B given that event A has occurred, Principle 4 is

$$P(A \text{ and } B) = P(A) * P(B|A) \qquad \text{for dependent events.}$$

Principle 5 is a restatement of Principle 4, namely

$$P(B|A) = \frac{P(A \text{ and } B)}{P(A)}.$$

This principle is a restatement of the second multiplication rule as a definition of *conditional* probability. Here we are introducing terminology not used by Laplace. But this is helpful in view of what is to come. Let us go back to the urn problem and make a list of all the "cases," or outcomes of the two draws. Since there are two white balls we will distinguish them by W1 and W2. The total cases are

Outcome	1	2	3	4	5	6
First draw	W1	W1	W2	W2	B	B
Second draw	W2	B	W1	B	W1	W2

We can condense these six equally likely outcomes into a two-way probability table, as follows:

	Second draw		
First draw	W	B	Total
W	2/6	2/6	4/6
B	1/6	1/6	2/6
Total	3/6	3/6	6/6

Now, Principle 5 says that the probability of white on the second draw, given that white was drawn on the first draw, is

$$P(\text{W on 2nd}|\text{W on 1st}) = \frac{P(\text{W on 1st and W on 2nd})}{P(\text{W on 1st})}$$

$$= \frac{2/6}{4/6} = 1/2.$$

Our examples of Principles 4 and 5 involved a sequence of events, and the conditional probability in each case was the probability of the second event given that the first event had occurred. Let us talk now about the probability of the first event given that the second has occurred. For example,

Suppose there are two urns, X and Y. Urn X contains two white and one black balls, while urn Y contains one white and two black balls. An urn is selected by tossing a coin, and from the urn selected a ball is drawn. The ball is black. What is the probability that urn X was selected?

The two urns are the antecedent conditions, or causes, which can give rise to the observation of a white or a black ball. In the probability posed above, we have observed an outcome (black ball) and are asked for the probability that urn X produced (caused) the black ball. The term "states" is used more than causes today, so we are asking for the probability that state X produced the black ball.

Using Principle 4 we can find the probabilities for all the possible sequences in the problem.

$$P(\text{X and W}) = (1/2) * (2/3) = 2/6,$$
$$P(\text{X and B}) = (1/2) * (1/3) = 1/6,$$
$$P(\text{Y and W}) = (1/2) * (1/3) = 1/6,$$
$$P(\text{Y and B}) = (1/2) * (2/3) = 2/6.$$

For display let us construct a two-way table, as follows:

State (S)	Event (E)		
	W	B	Total
Urn X	2/6	1/6	3/6
Urn Y	1/6	2/6	3/6
Total	3/6	3/6	6/6

Laplace's Principle 6 states that "the probability of the existence of any one (cause) ... is then a fraction whose numerator is the probability of the event resulting from this cause and whose denominator is the sum of similar probabilities relative to all the causes."

$$P(\text{Si}|\text{Ej}) = \frac{P(\text{Si and Ej})}{\sum P(\text{S and Ej})}.$$

Here, Si stands for a particular state and Ej stands for the event that occurred. The principle allows us to calculate the probability that the event (or evidence) that occurred was produced by a particular state.

For our example,

$$P(X|B) = \frac{P(X) * P(B|X)}{P(X) * P(B|X) + P(Y) * P(B|Y)}$$

$$= \frac{(1/2) * (1/3)}{(1/2) * (1/3) + (1/2) * (2/3)} = \frac{1/6}{3/6} = 1/3.$$

Principle 6 is known as Bayes' theorem after the Rev. Thomas Bayes (1702–1761). The paper containing the theorem was read posthumously to the Royal Society in 1763 by Richard Price (1723–1791), a noted mathematician. The process of reasoning from an event back to causes, or states, is called *inverse probability*, a term introduced by Laplace in 1773.

Principle 7 is embodied in Principle 6. It says that the probability of an event is the sum of the products of possible causes of the event and the conditional probabilities of the event given each cause. It simply elaborates how to get the probability called for in the denominator of Bayes' theorem.

$$P(Ej) = \sum [P(S \text{ and } Ej)] = \sum [P(S) * P(Ej|S)].$$

In our example, the denominator of the probability for urn X given a black ball is the probability of a black ball. The probability of a black ball is the sum of products of probabilities of states (urns) and probabilities of a black ball given each state.

$$\begin{aligned}
P(B) &= \sum [P(S) * P(B|S)] \\
&= P(X) * P(B|X) + P(Y) * P(B|Y) \\
&= (1/2) * (1/3) + (1/2) * (2/3) \\
&= 1/6 + 2/6 \\
&= 3/6.
\end{aligned}$$

Principle 9 concerns what Laplace called "mathematical hope," or "advantage," which we now call *expected value*. In Laplace's example Paul receives 2 francs if he throws heads on the first throw of a coin and 5 francs if heads does not occur until the second throw. We use V for value and EV for expected value and would calculate Paul's expected value as

$$\begin{aligned}
EV &= P(A) * (V|A) + P(B) * (V|B) \\
&= (1/2) * 2 \text{ francs} + (1/4) * 5 \text{ francs} \\
&= 1 \text{ franc} + 1.25 \text{ francs} \\
&= 2.25 \text{ francs}.
\end{aligned}$$

The advantage, or expected value, is 2.25 francs. "It is the sum which (Paul) ought to give in advance to that one who has given him this advantage; for, in order to maintain the equality of play, the throw ought to be equal to the advantage which it procures." (p. 20)

Laplace's Principle 9 is an elaboration of Principle 8. It says that when losses are involved we find the advantage (expected value) by subtracting the sum of products of losses and their probabilities from the sum of products of gains and their probabilities.

The tenth principle of Laplace involves what we now call utility and what Laplace called "moral hope." We postpone this topic to a later chapter.

Counting the Ways

In working probability problems it is useful to be able to count the number of ways in which certain events can happen. For example, the writer of a letter in *The American Statistician* compared the numbers of games played in the first 67 Baseball World Series with the distribution of lengths of a best-of-seven series that would be expected to occur between teams that were evenly matched. [7] The probability that a series between evenly matched teams lasts only four games is the probability that the losing team wins none of the first three games. This is the same as tossing three heads is a row in coin tossing.

P(series lasts four games) = (1/2)(1/2)(1/2) = 0.125.

We could also observe that there are eight possible outcomes of the first three games. There are two possible outcomes of the first game and, for each of these there are two possible outcomes of the second games, and for each of these four outcomes there are two outcomes for the third game. The number of distinct win–lose sequences for three games is, then, $2 \times 2 \times 2 = 8$, and they are equally likely because we are talking about equally matched teams. Since the sequences are equally likely, one-eighth is the probability of the three-loss sequence.

The probability of a five-game series is the probability that the losing team wins one of the first four games—just like throwing one head in four coin tosses. There are $2 \times 2 \times 2 \times 2 = 16$ distinct sequences for four tosses, and one head can occur in four ways, as follows:

H and T and T and T,

T and H and T and T,

T and T and H and T,

T and T and T and H.

So the probability of a five-game series is 4/16 = 0.25.

The probability of a six-game series is the probability that the losing team wins two of the first five games. For five games there are $2 \times 2 \times 2 \times 2 \times 2 = 32$ distinct win–lose sequences. In how many of these are there two

Table 5-1. Actual Lengths of First 67 World Series and
Probabilities for Evenly Matched Teams.

Number of games	Number of series	Relative frequency	Probability for evenly matched teams
4	12	0.179	0.1250
5	16	0.239	0.2500
6	15	0.224	0.3125
7	24	0.358	0.3125

losses? The answer is called the number of combinations of five things taken
two at a time. A formula for the number of combinations of n things taken r at
a time is

$$C_r^n = \frac{n(n - 1)\ldots(n - r + 1)}{1(2)\ldots(r)}.$$

For five things taken two at a time, $n = 5$ and $r = 2$, and

$$C_2^5 = \frac{5(4)}{1(2)} = 10,$$

and the probability that the team losing the World Series loses two of the first
five games (and the series lasts six games) is $10/32 = 0.3125$. For the series to
last seven games the losing team must win three of the first six games. There
are $2 \times 2 \times 2 \times 2 \times 2 \times 2 = 64$ win–lose sequences for six games and three
losses occur in

$$C_3^6 = \frac{6(5)(4)}{1(2)(3)} = 20$$

of these. The probability of a seven-game series is $20/64 = 0.3125$. The probabilities we have worked out and the actual record for the first 67 World Series
are given in Table 5-1.

We leave until later the question whether the differences between the
historical relative frequences and the probabilities for evenly matched teams
are large enough to conclude that World Series opponents were not evenly
matched.

Poker Hands

Poker hands are an interesting example of combinations. From a standard
card deck of 2, 3, . . . , 10, J, Q, K, Ace of clubs, diamonds, hearts, and spades,
you will be dealt five cards. What is the probability that you will be dealt a
straight flush, four-of-a-kind, etc.?

The number of different hands of five cards from 52 is the number of combinations of 52 things taken five at a time, or

$$C_5^{52} = \frac{52(51)(50)(49)(48)}{1(2)(3)(4)(5)} = 2{,}598{,}960.$$

How many of these will give rise to five cards in sequence of the same suit, remembering that Ace can count low or high? There are ten straights in each suit (ending with 5, 6, ..., K, Ace). With four suits, there are $4 \times 10 = 40$ ways of being dealt a straight flush.

Four-of-a-kind requires four of the same denomination and one other card. There are 13 denominations and 48 cards that can be dealt as the nonmatching card. So there are $13 \times 48 = 624$ ways of being dealt four-of-a-kind.

A full house requires three of one denomination and two of another. There are four ways of getting three of any one denomination and 13 denominations. Then, the pair must be in another denomination. There are 12 of these and six ways of pairing four cards of any denomination. The product is $13 \times 4 \times 12 \times 6 = 3744$ ways of getting a full house.

To be dealt a flush (five cards of the same suit) we must be dealt five of the 13 cards in the suit, and of course there are four possible suits.

$$4 * C_5^{13} = \frac{4(13)(12)(11)(10)(9)}{1(2)(3)(4)(5)} = 5148 \text{ ways}$$

from which we subtract 40 for the straight flushes above for 5108 ways.

There are ten possible straights (ending in 5, 6, 7, ..., Q, K, Ace) and each has $4 \times 4 \times 4 \times 4 \times 4$ ways of occurring, for $10 \times 1024 = 10{,}240$ ways. From this subtract the 40 straight flushes for 10,200 ways.

Three-of-a-kind requires one of four ways of getting three of the four cards of any one of 13 denominations. Any one of 48 other cards will be a different denomination from the triple, and any one of 44 cards will be a different denomination from the triple or the other nonmatching card. There are $48 \times 44/2$ nonmatching pairs that will fill the hand. Thus, there are $4 \times 13 \times 48 \times 44/2 = 54{,}912$ ways of getting three-of-a-kind.

By similar reasoning, two pair and one pair lead to the following numbers of ways.

$$C_2^4 * C_2^4 * C_2^{13} * 44 = 6 * 6 * 13 * 12 * 44/2 = 123{,}552,$$
$$13 * C_2^4 * 48 * 44 * 40/(1 * 2 * 3) = 109{,}240.$$

There are 1,302,500 remaining deals out of the 2,598,960 combinations of 52 cards five at a time. These hands have no value in poker. Dividing each number of ways by 2,598,960 yields the probabilities in Table 5-2 for poker hands.

Table 5-2. Probabilities of Poker Hands.

Hand	Probability	Hand	Probability
Straight flush	0.00002	Three-of-a-kind	0.02112
Four-of-a-kind	0.00024	Two pairs	0.04754
Full house	0.00144	One pair	0.42257
Flush	0.00197	Nothing	0.50116
Straight	0.00392	Total	1.00000

The Birthday Problem

There are 30 children in a classroom. What is the probability that at least two of them have the same birthday? This is a celebrated probability problem because most people grossly underestimate the probability. Actually the probability is better than 7 in 10 and the solution is easier than is imagined at first glance. The rationale is as follows:

The first child can have one of 365 birthdays. The chance that the second child *does not* have the same birthday is 364 out of 365. Given that the first two have different birthdays, the probability that the third has a different birthday is 363/365. And so it goes. Where n is the number of children, the probability of all n having different birthdays is

n	Probability
2	$\dfrac{364}{365}$
3	$\dfrac{364(363)}{365(365)}$
4	$\dfrac{364(363)(362)}{365(365)(365)}$
n	$\dfrac{364(363)(362)\ldots(365 - n + 1)}{(365)^{n-1}}$

To get the probability of a least two children having the same birthday, we subtract the results above from 1.0. A series like the one here is ideally suited to a computer. We put the birthday problem on our home computer, using a spread-sheet program, and obtained the results in Table 5-3 for 2–40 children. According to Slonim, the uninitiated will reason that with 30 people and around 360 possible birthdays the chance of a duplication in birthdays is about 30 in 360, or 1 in 12. [8] Now, the probability exceeds 7 in 10 because we must get 29 failures to match in a row. Laplace warned about the effect of many chances in a row, even when the individual chances are high.

Table 5-3. Probability that Two or More Have the Same Birthday as a Function of Size of Group.

n	Probability	n	Probability	n	Probability	n	Probability
2	0.0027	12	0.1944	22	0.5073	32	0.7750
3	0.0082	13	0.2231	23	0.5383	33	0.7953
4	0.0271	14	0.2529	24	0.5687	34	0.8144
5	0.0405	15	0.2836	25	0.5982	35	0.8322
6	0.0562	16	0.3150	26	0.6269	36	0.8487
7	0.0473	17	0.3469	27	0.6544	37	0.8641
8	0.0743	18	0.3791	28	0.6810	38	0.8782
9	0.1170	19	0.4114	29	0.7063	39	0.8912
10	0.1411	20	0.4437	30	0.7304	40	0.9032

Probability in Court

If probability is truly the scientific way to handle uncertainty, one would suppose that the courts would use it. In fact, the use of probabilistic evidence in court is a controversial subject. One element of controversy is whether probability has meaning when applied to past events. Some say probability should be restricted to future events and is meaningless when applied to past events—which either did or did not occur. Even where probabilistic evidence about past events has been allowed, it has sometimes been so conjectural that it has been thrown out on appeal.

In *People vs Risley* (214 N.Y. 75, 108 N.E. 200 [1915]) a mathematics professor was allowed to testify as to the probability that words allegedly inserted in a document in a fraud case were typed on the defendant's typewriter. The defendant was convicted but the appellate court held that the evidence was prejudicial and constituted "an attempt to draw a line between an assumed fact and a reasonable conclusion to an extent never recognized by this court." [9] One is reminded of the jibe that statistics is "... a mathematically precise way of drawing a straight line from an unwarranted assumption to a foregone conclusion."

Another mathematics professor fared no better in the most famous case involving probability, *People vs Collins* (69 Cal. 2nd at 325, 438 P.2d at 39, 66 Cal. Rptr. at 502). A robbery had been committed by a black man with a moustache and beard in the company of a white woman with blonde ponytails. Two persons answering the description were apprehended. The prosecutor used the professor's testimony to calculate, based on the joint occurrence of independent events, that there was but one chance in twelve million that another couple would possess the same characteristics as the defendants.

The appeals court wrote that "mathematics, a veritable sorcerer in our computerized society, while assisting the trier of fact in the search for truth must not cast a spell over him. We conclude that ... the defendant should not have his guilt determined by the odds and that he is entitled to a new trial." [10]

The problem here was not so much with probability as with how it was used. There was inadequate factual foundation about the constituent probabilities and their independence and, given this, an undue reliance on a distracting and irrelevant expert demonstration.

What seems to be the case is that courts will accept probabilistic evidence if a factual basis has been established, such as in occupational mortality and accident rates, but are unwilling to accept statistical speculation. Where individual guilt is in question, persuasion by the use of odds has been resisted.

Which Question Was That?

A perennial problem in survey research is getting answers to sensitive questions. While the amount of information that trained personal interviewers can get from respondents is often amazing, there are subjects that people are hesitant to respond to. To get truthful responses to subjects involving moral, sexual, or criminal acts is difficult. Respondents do not want to reveal these to an interviewer or to feel that their answers can be attributed to them in any way.

A way around this problem is known as the *randomized response* technique. In an article entitled "How to Get the Answer Without Being Sure You've Asked the Question," Cathy Campbell and Brian L. Joiner discuss several variations of the method. [11] In the simplest of these, a random event whose outcome is not known to the interviewer determines whether the respondent answers the question of interest or an unrelated question such as "Is your birthday on an even-numbered date?" The probability of a "yes" answer to the unrelated question must be known from underlying principles or from factual evidence like a recent census. In the case here the probability is 0.50.

Suppose the sensitive question is "Have you smoked pot in the last two weeks?" Whether a respondent is to answer this question will be determined by a random event. Suppose the event is picking an orange ball from a cannister which the respondent knows contains four orange and six green balls. The respondent shakes out a ball in a way that cannot be seen by the interviewer and returns it to the cannister. If the ball is orange the respondent is to reply to the sensitive question and if the ball is green is to reply to the question about birthdate. The interviewer does not know which question is being answered.

When all the returns are in the proportion answering "yes" to the sensitive question can be estimated. The estimated proportion of yes answers is

$$P = P(Orange) * P(Yes|Orange) + P(Green * P(Yes|Green),$$
$$P = 0.40 * P(Yes|Orange) + 0.60 * 0.50.$$

Everything is known here except P(Yes|Orange), which is the proportion of respondents answering the "pot" question who reply in the affirmative. Suppose the overall percentage of "yes" answers in 38. Then we can estimate

$$P(\text{Yes}|\text{Pot Question}) = (0.38 - 0.60 * 0.50)/0.40$$
$$= 0.20 \text{ or } 20\%.$$

Randomized response techniques have been successfully used in surveys on tax cheating, wife and child battering, and other socially sensitive or incriminating behaviors.

References

1. Helen M. Walker, *Studies in the History of Statistical Method*. Williams and Wilkins, Baltimore, MD, 1929, p. 7.
2. F. N. David, *Games, Gods, and Gambling*, Hafner, New York, 1962, p. 132.
3. Walker, op. cit., p. 12.
4. Arne Fisher, *The Mathematical Theory of Probabilities*, Macmillan, New York, 1923, p. 11.
5. Pierre Simon de Laplace, *A Philosophical Essay on Probabilities* (translated from the sixth French edition), Dover, New York, 1951.
6. Laplace, op. cit., p. iv.
7. Letter from R. E. Cavicehi, *The American Statistician*, **26**, No. 4 (October 1972), 55.
8. Morris J. Slonim, *Sampling in a Nutshell*, Simon and Schuster, New York, 1960, p. 10.
9. 214 N.Y. 75 at 87, quoted in Sharon Collins, "The Use of Social Research in the Courts," in *Knowledge and Policy: The Uncertain Connection*, National Research Council Study Project on Social Research and Development, National Academy of Sciences, Washington, DC, 1978, p. 155.
10. 68 Cal. 2nd 319 at 320, 438 P.2d at 33, quoted in Collins, op. cit., p. 156.
11. Cathy Campbell and Brian L. Joiner, "How to Get the Answer Without Being Sure You've Asked the Question," *The American Statistician*, **27**, No. 5 (December 1973), 229–231.

CHAPTER 6

Craps and Binomial

One of our illustrations of combinations in the last chapter was the number of ways that the baseball World Series could last four games, five games, six games, and seven games. Our interest there was to show how counting numbers of ways is part and parcel of probability calculations. That example led to a set of probabilities for the possible lengths of the series. When the events of concern in an uncertain situation are numerical and we have obtained the probabilities for all possible events, we have what is called a *probability distribution.*

A probability distribution is much like a frequency distribution. A frequency distribution shows actual results, like the lengths of the first 67 World Series. A probability distribution shows the *relative frequencies* of outcomes to be expected in a large number of observations given certain assumptions.

We may want to compare actual observations with calculated probabilities as a test of these assumtions—as in the World Series example. In other cases, as in gambling devices, casino operators are very concerned to have systems that meet the usual assumptions of evenly balanced dice, honest dealers, etc. Then an operation will be predictable, honest, and profitable.

Probabilities for Craps

Figure 6-1 presents graphically the probability distribution for the sum of spots on two dice. It is the first step in working out the probability that the roller wins a game of craps. The example illustrates how probabilities are used to find the probability of a complex outcome from the simple outcomes that comprise it—using the principles of probability calculations. The rules of the

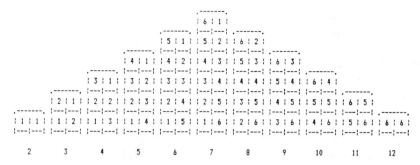

Figure 6-1. Outcomes of a Throw of Two Dice.

game of craps are:

> The roller throws two dice, and loses if 2, 3, or 12 is rolled. If 7 or 11 is rolled, the roller wins. If none of the above happens, the number rolled is called point. The roller then continues to roll, and wins if point is rolled before a 7 is rolled.

The probability for the roller winning is calculated in Table 6-1, which deserves some comment. The first column, P(S), shows the probabilities for the sums (S) on throwing two dice. These can be verified by counting the numbers of equally likely ways in Figure 6-1, or by calculations. For example, the probability of throwing a 3 is P(1 and 2) + P(2 and 1) = (1/6)(1/6) + (1/6)(1/6) = 2/36. The probabilities are expressed in 36ths, and then repeated as decimal fractions.

Table 6-1. Probability of Roller Winning a Game of Craps.

S	P(S)	P(Point|S)	P(S) * P(Point|S)
2	1/36 = 0.0277778		
3	2/36 = 0.0555556		
4	3/36 = 0.0833333	3/(3 + 6) = 0.333333	0.0277777
5	4/36 = 0.1111111	4/(4 + 6) = 0.400000	0.0444444
6	5/36 = 0.1388889	5/(5 + 6) = 0.454545	0.0631313
7	6/36 = 0.1666667		
8	5/36 = 0.1388889	5/(5 + 6) = 0.454545	0.0631313
9	4/36 = 0.1111111	4/(4 + 6) = 0.400000	0.0444444
10	3/36 = 0.0833333	3/(3 + 6) = 0.333333	0.0277777
11	2/36 = 0.0555556		
12	1/36 = 0.0277778		
Total	36/36 1.0000000		0.2707071

P(Roller wins on first throw) = P(7) + P(11) = 0.2222222
P(Roller wins after first throw) =   0.2707071
P(Roller wins a game of craps) =   0.4929293

There are two ways the roller can win—by winning on the first roll, or by winning subsequently. At this point we can find the probability that the roller wins on the first roll, which is P(7 or 11) = 0.2222. We now need the probability of the roller winning subsequent to the first roll.

The numbers 4, 5, 6 and 8, 9, 10 are possible points. If 4 is point, the chances of rolling point before 7 are 3 to 6, since there are three chances in 36 for the 3 and six chances in 36 for the 7. Probability is the ratio of favorable to total chances, or 3/9 for rerolling point before 7. Under P(Point|S) are shown the probabilities of rolling point before a 7 for the possible values of point. Then P(S) for each possible point is multipled by the probability of rerolling point before 7 (for that point) to get the probability of winning the game in that particular way. Then these products are added because they represent probabilities for the several ways of winning the game after the first roll—rerolling point before a 7. This sum, 0.2707, is added to 0.2222, the probability of the roller winning on the first roll, to get the probability that the roller wins the game, 0.4929.

The Binomial Distribution

In 1974 a $1,250,000 settlement for the plaintiffs was awarded in a class action suit against the Hacienda Hotel of Las Vegas, Nevada. The hotel had placed advertisements in the *Los Angeles Times* for a package plan vacation much as follows. [1]

FIESTA HOLIDAY
2 DAYS/1 NIGHT FOR ONE OR TWO PERSONS
You pay $15
WE GIVE YOU $10 in FREE play chips!
Your NET Cost is $5.

Upon arrival at the hotel, customers learn that the "FREE play chips" are to be used during the course of gambling. If the customer wins when playing such a chip, the chip must be given up and the customer wins $1. If the customer loses when playing the chip, the chip is given up and the customer receives nothing.

Let us suppose the customer is playing craps and is betting "the line," that is, is betting that the roller wins the game. Any time the customer bets the special chip, he/she makes a bet with the following expectation.

0.493 Probability of winning $1 = $0.493,
0.507 Probability of winning $0 = $ –0–,

Expected value   = $0.493.

In betting the line in craps (probably the fairest bet available) each chip is worth 49.3 cents. The ten chips are worth $4.93, or not quite $5, so the "NET" cost to the customer is right around $10 rather than $5. That was the basis of the award. If all the customers played craps they would wind up paying $15 − $4.93 = $10.07 on the average. Let us pursue the matter further and ask about variation from this average.

We will need to know the probability distribution of numbers of wins in ten trials of playing a free chip. The individual trials occur under conditions of constant probability of a success and they are independent of one another. Under these conditions they are known as Bernoulli trials, after the Swiss mathematician, Jacques (Jacob) Bernoulli (1654–1705). For ten trials the possible numbers of wins are 0, 1, 2, ..., 10, and the probabilities can be found from

$$P(r) = C_r^n * P^r * (1 - P)^{n-r},$$

where n is the number of trials (ten in our case), r is the number of wins (successes), and P is the constant probability of success on any trial (0.493 in our case).

The probability distribution of number of successes in n Bernoulli trails is called the *binomial* distribution. To specify a binomial distribution we need to know n (the number of trials) and P (the probability of success on any trial). Probabilities for the binomial distribution for $n = 10$ and $P = 0.493$ are given in Table 6-2.

Table 6-2 also shows Pascal's triangle (after Blaise Pascal, (1623–1662), which gives the combinations of n things r at a time that are required in the binomial formula. Pascal's triangle works as follows:

Write down

```
     1
   1   1.
```

Then, write 1 at the extremes and in between numbers in the preceding row write their sum. So, with the third row we have

```
       1
     1   1
   1   2   1,
```

and adding the fourth row,

```
         1
       1   1
     1   2   1
   1   3   3   1,
```

and so on. The fourth row contains the binomial coefficients for three trials ($n = 3$). We carried Pascal's triangle out to show the coefficients for $n = 10$ (trials). Of course, one can always go back to the combinations formula shown earlier.

The binomial distribution tells us that there is a 4.77 percent chance that a

Table 6-2. Pascal's Triangle and Binomial for $n = 10$, $P = 0.493$.

													r	P(r)
									1				0	0.0011
								1						
							1		9			10	1	0.0109
						1		8				45	2	0.0477
					1		7		36					
				1		6		28				120	3	0.1238
			1		5		21		84					
		1		4		15		56				210	4	0.2107
	1		3		10		35		126					
1		2		6		20		70				256	5	0.2498
	1		3		10		35		126					
		1		4		15		56				210	6	0.1922
			1		5		21		84					
				1		6		28				120	7	0.1107
					1		7		36					
						1		8				45	8	0.0404
							1		9			10	9	0.0087
								1						
									1				10	0.0008

Sum   1   2   4   8   16   32   64   128   256   612   1024

customer will win with only two of the ten free chips, and a 4.04 percent chance of winning with eight of the ten chips. In these cases the customer would save $2 and $8, respectively, making the net cost of the weekend package $13 in the one case and $7 in the other. The chances of winning on all of the free chips is 0.0008. There is not quite one chance in a thousand that the net cost will be $5, as advertised, even when customers play the fairest game available.

Inference from the Binomial

W. J. Youden (1900–1971), long-time statistician with the U.S. Bureau of Standards, wrote

> There is a great deal of difference between calculating the probability of getting a particular throw from a pair of dice—known or assumed to be absolutely true— and the problem of looking at a collection of such throws and then hazarding an opinion, or probability statement, about the dice. [2]

This is the position we appear to be in at this stage in our journey. We have seen how to calculate probabilities of certain outcomes in dice throwing and coin tossing, but these assume perfectly balanced dice or coins for which we can say that the probability for any single throw is 1/2 for heads and 1/6 for an Ace. With such values and the principles for probability calculations, the probabilities for complex events are deductive matters.

We are in a position now to look briefly at the use of probability distributions in *induction*, or statistical inference. Our example will be from coin-tossing rather than dice. It is

An investigator comes into possession of a particular Eisenhower silver dollar. After tossing the dollar eight times, three heads have been observed. What can be said about the probability of heads for the coin?

Table 6-3 shows some binomial distributions for $n = 8$. Each of these shows probabilities of the outcomes $r = 1, 2, \ldots, 8$ heads *if the underlying probability for the Bernoulli trials was as shown*. How can we use this information to help us?

The actual result was $r = 3$ (heads) in $n = 8$ (tosses). We do not know which of the binomials our sample result belongs to. It might be eight Bernoulli trials from a population in which P (the long run proportion of heads) is almost any of the values of P shown. What would a good estimate be?

A characteristic of any binomial distribution is that if the values of r/n, the proportion of successes for the sample trials, are calculated and then averaged (weighted by the probabilities), the result will be P, the true probability for the process. This means that if we were to take repeated samples of n trials, the average sample proportion, r/n, would approach the true probability.

The reasoning above supports the use of r/n as the *estimate* of the true probability. Using r/n as an estimate of P is an *unbiased* procedure. This means that the average of repeated estimates will tend to equal the true probability. The estimate r/n in our sample is $3/8 = 0.375$.

Another rationale can be appreciated by scanning across the row for $r = 3$, reproduced below the main table. There we see the probability of the observed result *for different values of P*. These are the *likelihoods* of the sample result. Note that the likelihoods reach a maximum at 0.40. The probability of $r = 3$ given $P = 0.40$ is 0.2787, higher than for any other P value. From this we could call 0.40 the *maximum likelihood* estimate of the probability of a head on the Eisenhower dollar in question. In general, the maximum likelihood estimate for P in a Bernoulli process is

$$\frac{r + 1}{n + 2}.$$

In our case, $(3 + 1)/(8 + 2) = 0.40$.

The relationship above is called Laplace's *law of succession*. At one extreme it says if you have no observations concerning a Bernoulli process, your best estimate of P is 1/2. At the other extreme, with r and n growing large, the law converges to r/n, the sample proportion.

Table 6-3. Binomial Distributions for $n = 8$.

r \ P	0.05	0.10	0.15	0.20	0.25	0.30	0.35	0.40	0.45	0.50	0.55	0.60	0.65	0.70	0.75	0.80	0.85	0.90	0.95
0	0.6634	0.4305	0.2725	0.1678	0.1001	0.0576	0.0319	0.0168	0.0084	0.0039	0.0017	0.0007	0.0002	0.0001	0.0000	0.0000	0.0000	0.0000	0.0000
1	0.2793	0.3826	0.3847	0.3355	0.2670	0.1977	0.1373	0.0896	0.0548	0.0313	0.0164	0.0079	0.0033	0.0012	0.0004	0.0001	0.0000	0.0000	0.0000
2	0.0459	0.1329	0.2121	0.2621	0.2781	0.2647	0.2310	0.1366	0.1401	0.0977	0.0623	0.0369	0.0194	0.0089	0.0034	0.0010	0.0002	0.0000	0.0000
3	0.0054	0.0331	0.0839	0.1468	0.2076	0.2541	0.2786	0.2787	0.2568	0.2188	0.1719	0.1239	0.0808	0.0467	0.0231	0.0092	0.0026	0.0004	0.0000
4	0.0004	0.0046	0.0185	0.0459	0.0865	0.1361	0.1875	0.2322	0.2627	0.2734	0.2627	0.2322	0.1875	0.1361	0.0865	0.0459	0.0185	0.0046	0.0004
5	0.0000	0.0004	0.0026	0.0092	0.0231	0.0467	0.0808	0.1239	0.1719	0.2188	0.2568	0.2787	0.2786	0.2541	0.2076	0.1468	0.0839	0.0331	0.0054
6	0.0000	0.0000	0.0002	0.0010	0.0034	0.0089	0.0194	0.0369	0.0628	0.0977	0.1401	0.1866	0.2310	0.2647	0.2781	0.2621	0.2121	0.1329	0.0459
7	0.0000	0.0000	0.0000	0.0001	0.0004	0.0012	0.0033	0.0079	0.0164	0.0313	0.0548	0.0896	0.1373	0.1977	0.2670	0.3355	0.3847	0.3826	0.2793
8	0.0000	0.0000	0.0000	0.0000	0.0000	0.0001	0.0002	0.0007	0.0017	0.0039	0.0084	0.0168	0.0319	0.0576	0.1001	0.1678	0.2725	0.4305	0.6634

r \ P	0.05	0.10	0.15	0.20	0.25	0.30	0.35	0.40	0.45	0.50	0.55	0.60	0.65	0.70	0.75	0.80	0.85	0.90	0.95
0			0.2725										0.0002						
1			0.3847										0.0033						
2			0.2121										0.0194						
3	0.0054	0.0331	0.0839	0.1468	0.2076	0.2541	0.2786	0.2787	0.2568	0.2188	0.1719	0.1239	0.0808	0.0467	0.0231	0.0092	0.0026	0.0004	0.0000
4			0.0185										0.1875						
5			0.0026										0.2786						
6			0.0002										0.2310						
7			0.0000										0.1373						
8			0.0000										0.0319						

Interval Estimates

Both of the estimates suggested above are what are called *point* estimates. We want to use the sample value to suggest some single number which is in some sense a good estimate. The criteria of unbiasedness and maximum likelihood are examples of criteria for good point estimates. But while they have these good qualities, we have no measure of their reliability. In fact, no single-value estimate is one in which you can put much trust.

To get a measure of trustworthiness we have to make an interval estimate. We have to give a range of values which we estimate to contain the probability, P. To do this in a way that leads to a measure of confidence in the procedure, consider the following.

Find a value of P so low that the probability of r equal to or exceeding the observed number is equal to some small value, $A/2$.

$$P[R \geq r | P(\text{low})] = A/2.$$

Find a value of P so high that the probability of r equal to or less than the observed number is equal to some small value, $A/2$.

$$P[R \leq r | P(\text{high})] = A/2.$$

Repeated estimates from a Bernoulli process made in this way will contain P in $1 - A$ proportion of the cases in the long run. The reason is that the probability that the observed r lies in the central $1 - A$ proportion of the binomial distribution for r, *whatever P is*, is exactly $1 - A$. For this reason the range

$$P(\text{low}) \text{ to } P(\text{high})$$

is an interval estimate of P in which we can place a measure of confidence, $1 - A$.

This is a concept at once simple and subtle. Suppose we set A equal to about 0.20. Remember that we observed three heads. We first look for a binomial in which $P(R \geq 3)$ is around 0.10. We get close with the binomial with $P = 0.15$, in which this probability is $0.0839 + 0.0185 + 0.0026 + 0.0002 = 0.1052$. If we had binomial distributions at intervals of 0.01 we could get closer. Next, we look for a binomial with a high value of P such that $P(R \leq 3)$ is around 0.10. We find this at $P = 0.65$ where $P(R \leq 3) = 0.0808 + 0.0194 + 0.0033 + 0.0002 = 0.1055$. The range

$$P(\text{low}) \text{ to } P(\text{high}) = 0.15 \text{ to } 0.65$$

is an interval estimate which we have $1 - (0.1052 + 0.1055) = 0.79$ or 79 percent confidence is correct. We do not know whether the estimate is correct in this case, but we are using a method that will produce correct estimates in 80 percent of the instances when we use it.

References

1. Letter from A. R. Sampson in *The American Statistician*, **28**, No. 2 (May 1974), 76.
2. W. J. Youden, "How to Pick a Winner," *Industrial and Engineering Chemistry*, **50**, No. 6 (June 1958), 81A.

CHAPTER 7

Horsekicks, Deadly Quarrels, and Flying Bombs

In his 1951 book, *Facts from Figures*, M. J. Moroney has a chapter entitled "Goals, Floods, and Horsekicks." [1] One can only conclude that violence has increased in the world when we start with horsekicks and go on to deadly quarrels and flying bombs. All of these are examples of a distribution carrying the name of a French mathematician, Simeon D. Poisson (1781–1840).

The Poisson Distribution

The examples (shown in Table 7-1) are all historical ones cited in a text by Griffin. [2] The first is data from a study by Ladislaus von Bortkiewicz (1868–1931) on deaths to soldiers in the Prussian army from kicks by army mules. Bortkiewicz studied 280 corps-years of exposure to this unhappy phenomenon. In 144 of these he observed no deaths, in 91 corps-years one death was observed, and so forth.

The second example is from a study by Lewis W. Richardson entitled "The Statistics of Deadly Quarrels." Richardson noted the number of outbreaks of war in each of the years from 1500 to 1931, a total of 432 years. In 223 of these no war broke out, in 142 of the years one war broke out in the world, and so on.

The final example is the pattern of impacts of German V1 flying bombs on London during World War II. Investigators divided the area of London (12 × 12 kilometers) into 576 $\frac{1}{4}$-square-kilometer units. Then they recorded in how many of these units no bombs impacted, one bomb, two bombs, and so on.

All three are examples of accidental occurrences where exposure to the

Table 7-1. Poisson Examples.

Occurrences	Deaths from horsekicks from army mules			Outbreaks of war 1500–1931			Impacts of V1 flying bombs on ¼-kilometer areas in London		
	Number of corps-years	Poisson distribution		Number of years	Poisson distribution		Number of areas	Poisson distribution	
		Probability	Number		Probability	Number		Probability	Number
0	144	0.497	139.0	223	0.501	216.2	229	0.394	226.7
1	91	0.348	97.3	142	0.346	149.7	211	0.367	211.4
2	32	0.122	34.1	48	0.120	51.8	93	0.171	98.5
3	11	0.028	8.0	15	0.028	12.0	35	0.053	30.6
4	2	0.005	1.4	4	0.005	2.1	7	0.012	7.1
5		0.001	0.2		0.001	0.3		0.002	1.3
6									0.2
7							1		
	280		280.0	432		432.0	576		576.0
Mean	0.700			0.692			0.932		

accident is constant over all the units studied. The units are the corps-years, the calendar years, and the $\frac{1}{4}$-kilometer areas. The risk from mulekicks is essentially the same over all army corps and years; perhaps the risk of an outbreak of war has been constant since 1500; and it was known that the German V1 bombs had only a rudimentary guidance system sufficient to aim the bombs only generally at London.

Poisson's contribution was to develop the mathematics for the probability distribution that prevails when these conditions of constant exposure and independence among units of exposure prevail. The equation of the Poisson distribution is

$$P(r) = e^{-\mu}\left(\frac{\mu^r}{r!}\right),$$

where $r! = r(r-1)(r-2)\ldots(1)$ and $0! = 1$.

Here, e is the constant, 2.71828, and μ (the lowercase Greek letter, "mu"), which stands for mean, is the average number of occurrences per unit of exposure. In Table 7-1 we have calculated the mean for each distribution. They are 0.70 deaths per corps-year, 0.692 outbreaks per year, and 0.932 hits per area. Of course, these are only estimates for the long-run causal systems producing the occurrences in each case. Given a value for the mean, getting probabilities from Poisson's formula is just calculation.

The calculations are easily carried out on a mini-computer with a spreadsheet program. The reason is that the successive values of the right-hand term in parentheses in the formula can be obtained by an iterative process. That is,

$$\mu^0/0! = 1/1 = 1,$$
$$\mu^1/1! = 1 * \mu,$$
$$\mu^2/2! = (1 * \mu) * \mu/2,$$
$$\mu^3/3! = (1 * \mu * \mu/2) * \mu/3,$$

etc.

Having obtained the probabilities, they are translated into expected numbers of occurrences for comparison with the actual distribution pattern. Notice that the pattern produced by the Poisson formula is in each case close to the actual pattern. Apparently the conditions of equality and independence of exposure did prevail in these cases. Later, we will show you a rigorous way to test that question.

The Poisson distribution has proved to be applicable to a wide variety of situations. Occurrences of demand for spare parts and for individual sizes and varieties of assortments has often shown the Poisson pattern, as has the occurrence of defects in materials, and failures in systems. In our examples, where the occurrences per unit were less than one, the Poisson pattern is highly skewed to the right. As mean occurrences per unit of exposure increase, the skewness decreases, but is always present to a degree.

The Exponential Distribution

It might occur to you to try to look at the situations in these examples in another way. Rather than looking at the number of years with none, one, two, etc. outbreaks of war, why not look at the time between outbreaks. If there are 0.692 outbreaks per year, the average time between outbreaks must be $1/0.692 = 1.445$ years. Similarly, the time to next death by mulekick will be measured in corps-years, and what would interest us in the flying bomb example is the average distance between hits. This will be the square root of $1/0.932$, or 1.036 half-kilometers = 0.52 kilometer.

The time or distance to next occurrence is a *continuous* distribution, while the number of occurrences in time or distance units was a simple count. Consequently, instead of probabilities for exact values of the variable we have probabilities within intervals of values. In calculus terms, the probability within an interval of the variable is obtained by *integrating* an equation for frequency density over a range corresponding to the boundaries of the interval.

The distribution of times or distances to next occurrence when the Poisson conditions prevail is called the *exponential* distribution. These distributions have the density pattern

$$f(x) = \mu * e^{-\mu x}$$

and cumulative probabilities are given by

$$P(X \leq x) = 1 - e^{-\mu x},$$

where μ is the mean number of occurrences per unit of exposure, i.e., the Poisson mean.

In Table 7-2 we show the cumulative probabilities for the three examples. For example, the probability that the next outbreak of war will occur within a year is 0.499. The probability that the next flying bomb will impact within $\frac{1}{2}$ kilometer (root 1/4) is 0.606, and within 1 kilometer (root 4/4) is 0.976.

Exponential distributions all have continuously declining density, with maximum density at zero. When the mean interval to next occurrence $(1/\mu)$ is low, the density declines rapidly. For distributions with longer average interval to next occurrence, the initial density is not as high and the densities decline less rapidly.

Figure 7-1 shows Poisson distributions with mean occurrences per interval of 2.0 and 4.0. Also shown are the related exponential distributions of length of interval to next occurrence. These are shown with probabilities plotted in class intervals of 0.10 width. The Poisson with a mean of 4.0 is less skewed than the Poisson with a mean of 2.0. Poissons with increasing means show decreasing skewness. The exponential that is paired with the first Poisson has a mean of $\frac{1}{4} = 0.25$, while the second exponential distribution has a mean of $\frac{1}{2} = 0.50$. The first one illustrates the higher initial density and sharper decline in densities that we spoke of above.

Table 7-2. Exponential Distributions.

Time to next death (corps-years)		Time to next outbreak (years)		Distance (squared) to next hit (sq. km/4)	
Equal or less than	Cumulative probability	Equal or less than	Cumulative probability	Equal or less than	Cumulative probability
0.0	0	0.0	0	0.0	0
0.5	0.2953	0.5	0.2925	0.5	0.3726
1.0	0.5034	1.0	0.4994	1.0	0.6063
1.5	0.6500	1.5	0.6459	1.5	0.753
2.0	0.7534	2.0	0.7494	2.0	0.845
2.5	0.8262	2.5	0.8228	2.5	0.9028
3.0	0.8775	3.0	0.8746	3.0	0.9390
3.5	0.9137	3.5	0.9113	3.5	0.9617
4.0	0.9392	4.0	0.9372	4.0	0.9760
4.5	0.9571	4.5	0.9556	4.5	0.9849
5.0	0.9698	5.0	0.9686	5.0	0.9905

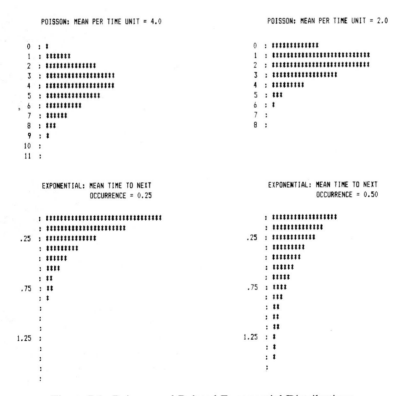

Figure 7-1. Poisson and Related Exponential Distributions.

A Problem of Authorship

An example of use of the Poisson distribution comes from an investigation by Frederick Mosteller and David L. Wallace into the likely authorship of some of the *Federalist Papers*. [3] The 85 *Federalist* papers were written by Alexander Hamilton, John Jay, and James Madison. They are an important source of knowledge of the intentions and philosophies of the framers of the Constitution of the United States. Historians agreed that Madison wrote 14 of the papers, Hamilton wrote 51, and Jay 5, but were uncertain about 15 of them. Of these 15, 12 were in dispute as between Hamilton and Madison. Both of these men wrote in what was called the *Spectator* style (complicated and oratorical), so that from the literary standpoint they were difficult to tell apart.

Mosteller and Wallace undertook an analysis of word frequencies in an effort to settle the question for each of the disputed 12 papers. The study was extensive and we give just a taste of the methodology. They examined known writings of Hamilton and Madison to find key words whose rate of usage varied between the two authors. Noncontextual words like although, enough, by, also, upon, are good candidates because they occur with some frequency in almost any writing and do not depend on subject matter.

A good example is the word *upon*. *Upon* occurs at a rate of about 6.5 per 2000 words in the writing of Hamilton, but only at a rate of 0.5 per 2000 words in Madison's prose. In *Federalist* Papers 54 and 55, each containing about 2000 words, *upon* occurs not at all in the first and twice in the second. Assuming an underlying Poisson process, one can ask what is the probability of each event if the writing was a sample of Hamilton's prose on the one hand and if it came from Madison's pen on the other.

For Paper 54 we calculate

$$P(R = 0|\text{Hamilton}) = e^{-6.5}\left(\frac{6.5^0}{1}\right) = 0.0015,$$

$$P(R = 0|\text{Madison}) = e^{-0.5}\left(\frac{0.5^0}{1}\right) = 0.6065.$$

And for Paper 55

$$P(R = 2|\text{Hamilton}) = e^{-6.5}\left(\frac{6.5^2}{2}\right) = 0.0318,$$

$$P(R = 2|\text{Madison}) = e^{-0.5}\left(\frac{0.5^2}{2}\right) = 0.0758.$$

We have calculated the likelihood of each event first given that Hamilton was the author and then given that Madison was the author. Hamilton and Madison are the two possible "causes" or "states" that could have given rise to the event. We see that the outcome in Paper 54 is much more likely given

Madison than Hamilton, and that the outcome in Paper 55 is somewhat more likely given Madison than Hamilton.

Many more words were analyzed and all the evidence weighed together. The investigators concluded that the evidence favored Madison overwhelmingly in 11 of the 12 disputed papers and substantially in the twelfth (Paper 55).

Another historical case of disputed authorship concerns plays carrying the name of William Shakespeare. At times claims have been made that Francis Bacon or Christopher Marlowe authored some of them. In a 1901 paper T. C. Mendenhall (1841–1924) reported on the frequency distributions of word lengths (letters per word) of various authors. He found the distribution for Shakespeare and Francis Bacon quite different, but he writes of Christopher Marlowe that "in the characteristic curve of his plays Marlowe agrees with Shakespeare about as well as Shakespeare agrees with himself." [4]

References

1. M. J. Moroney, *Facts from Figures*, 3rd ed., Penguin, Baltimore, MD, 1956.
2. J. Griffin, *Statistics: Methods and Applications*, Holt, Rinehart, and Winston, New York, 1962.
3. F. Mosteller and D. L. Wallace, *Inference and Disputed Authorship: The Federalist*, Addison-Wesley, Reading, MA, 1964.
4. Quoted in C. B. Williams, "A Note on an Early Statistical Study of Literary Style," *Biometrika*, **43** (1956), 248–256.

CHAPTER 8

Normal Distribution

Abraham De Moivre (1667–1754), who was born in France but lived most of his adult life in England, has the largest claim to the discovery of the normal distribution. The Frenchman Laplace and the German Gauss used the normal distribution especially in astronomic and geodesic measurements, and the Belgian Quetelet first applied it extensively to social statistics. [1]

Our introduction draws on data generated by a device called a *quincunx*, or probability board. The design of this board is shown in Figure 8-1. One-hundred small steel balls, about the size of beebee shot, are released through a gate at the top of the board into a maze of pins. The pins deflect the course of the balls on their journey to the 20 numbered slots at the bottom of the board. Without the pins, the balls would tend by gravity alone to fall between the 10 and 11 slots. But the pins deflect the balls to the right or left depending on exactly how they strike. Here we have a simulation of a constant cause (gravity) tending to produce results of a certain magnitude, with the addition of a large number of independent factors tending to produce deviations from the effects of the constant cause. These factors are called collectively *chance*.

Five series of ten runs each (five runs of 1000 balls) were run through the board. The results are shown in cumulative fashion in Figure 8-2. The first panel shows the outcomes of the first 1000 balls, the second shows the addition of the next 1000, and so on. The last panel shows the results of all 5000 balls. What we see is the build-up of the characteristic bell-shaped, normal distribution pattern. Minor irregularities that show in the first series of trials tend to be swamped eventually by the prevailing constant cause system. The chance factors tend to produce a pattern of symmetrical deviations from the central tendency in which small deviations are frequent and larger deviations successively less frequent. As Francis Galton wrote:

> I know of scarcely anything so apt to impress the imagination as the wonderful form of cosmic order expressed by the 'Law of Frequency of Error.' The law

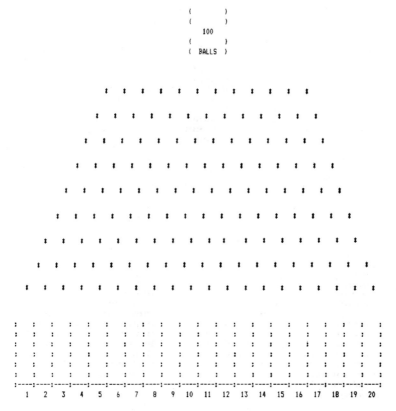

Figure 8-1. Probability Board.

would have been personified by the Greeks and deified, if they had known it. It reigns with serenity and complete self-effacement amidst the wildest confusion. The huger the mob and the greater the apparent anarchy, the more perfect its sway. It is the supreme law of unreason. Whenever a large sample of chaotic elements are taken in hand and marshalled in the order of their magnitude, an unsuspected and most beautiful form of regularity proves to have been latent all along. [2]

W. J. Youden (1900–1971), consultant for the U.S. Bureau of Standards, whose hobby was typography, composed this: [3]

THE

NORMAL

LAW OF ERROR

STANDS OUT IN THE

EXPERIENCE OF MANKIND

AS ONE OF THE BROADEST

GENERALIZATIONS OF NATURAL

PHILOSOPHY * IT SERVES AS THE

GUIDING INSTRUMENT IN RESEARCHES

IN THE PHYSICAL AND SOCIAL SCIENCES AND

IN MEDICINE AGRICULTURAL AND ENGINEERING

IT IS AN INDISPENSABLE TOOL FOR THE ANALYSIS AND THE

INTERPRETATION OF BASIC DATA OBTAINED BY OBSERVATION AND EXPERIMENT

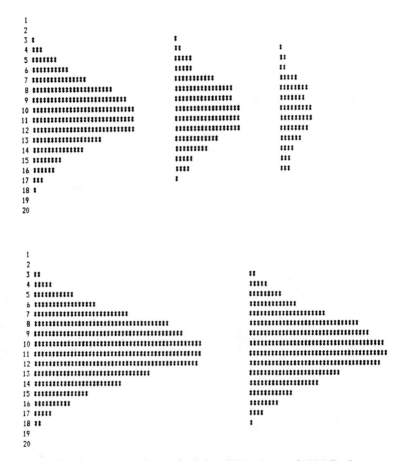

Figure 8-2. Cumulative Results of Five Runs of 1000 Each.

Figure 8-3 shows the form that would be produced by the probability board in the long run. It is a normal distribution with a mean of 1010.5 (we have added 1000 as we explain below) and a standard deviation of 3.0.

We want you to regard the distribution as representing the long-run performance of a container filling machine which is set to fill 1.0 liter (1000 ml) containers. It is known that the tolerance of the machine is such that the standard deviation of amounts filled at a fixed setting is 3.0 ml. The machine is set to fill containers with 1010.5 ml. The symbols for the long-run mean and standard deviation of a process are

$$\text{Mean:} \quad \mu = 1010.5,$$

$$\text{Standard deviation:} \quad \sigma = 3.0.$$

In Figure 8-3 you can see the reason for the setting the mean fill at 1010.5. We

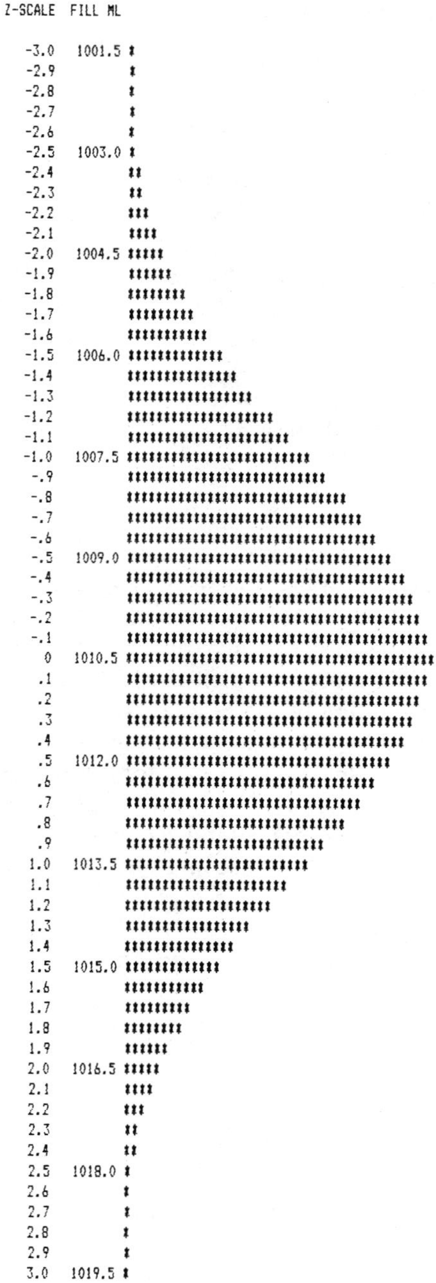

```
Z-SCALE FILL ML

 -3.0   1001.5 ‡
 -2.9          ‡
 -2.8          ‡
 -2.7          ‡
 -2.6          ‡
 -2.5   1003.0 ‡
 -2.4          ‡‡
 -2.3          ‡‡
 -2.2          ‡‡‡
 -2.1          ‡‡‡‡
 -2.0   1004.5 ‡‡‡‡‡
 -1.9          ‡‡‡‡‡‡
 -1.8          ‡‡‡‡‡‡‡
 -1.7          ‡‡‡‡‡‡‡‡
 -1.6          ‡‡‡‡‡‡‡‡‡
 -1.5   1006.0 ‡‡‡‡‡‡‡‡‡‡‡
 -1.4          ‡‡‡‡‡‡‡‡‡‡‡‡‡
 -1.3          ‡‡‡‡‡‡‡‡‡‡‡‡‡‡‡
 -1.2          ‡‡‡‡‡‡‡‡‡‡‡‡‡‡‡‡‡
 -1.1          ‡‡‡‡‡‡‡‡‡‡‡‡‡‡‡‡‡‡‡
 -1.0   1007.5 ‡‡‡‡‡‡‡‡‡‡‡‡‡‡‡‡‡‡‡‡‡
  -.9          ‡‡‡‡‡‡‡‡‡‡‡‡‡‡‡‡‡‡‡‡‡‡‡
  -.8          ‡‡‡‡‡‡‡‡‡‡‡‡‡‡‡‡‡‡‡‡‡‡‡‡‡‡
  -.7          ‡‡‡‡‡‡‡‡‡‡‡‡‡‡‡‡‡‡‡‡‡‡‡‡‡‡‡
  -.6          ‡‡‡‡‡‡‡‡‡‡‡‡‡‡‡‡‡‡‡‡‡‡‡‡‡‡‡‡
  -.5   1009.0 ‡‡‡‡‡‡‡‡‡‡‡‡‡‡‡‡‡‡‡‡‡‡‡‡‡‡‡‡‡‡
  -.4          ‡‡‡‡‡‡‡‡‡‡‡‡‡‡‡‡‡‡‡‡‡‡‡‡‡‡‡‡‡‡‡
  -.3          ‡‡‡‡‡‡‡‡‡‡‡‡‡‡‡‡‡‡‡‡‡‡‡‡‡‡‡‡‡‡‡‡
  -.2          ‡‡‡‡‡‡‡‡‡‡‡‡‡‡‡‡‡‡‡‡‡‡‡‡‡‡‡‡‡‡‡‡‡
  -.1          ‡‡‡‡‡‡‡‡‡‡‡‡‡‡‡‡‡‡‡‡‡‡‡‡‡‡‡‡‡‡‡‡‡‡
   0    1010.5 ‡‡‡‡‡‡‡‡‡‡‡‡‡‡‡‡‡‡‡‡‡‡‡‡‡‡‡‡‡‡‡‡‡‡‡
   .1          ‡‡‡‡‡‡‡‡‡‡‡‡‡‡‡‡‡‡‡‡‡‡‡‡‡‡‡‡‡‡‡‡‡‡
   .2          ‡‡‡‡‡‡‡‡‡‡‡‡‡‡‡‡‡‡‡‡‡‡‡‡‡‡‡‡‡‡‡‡‡
   .3          ‡‡‡‡‡‡‡‡‡‡‡‡‡‡‡‡‡‡‡‡‡‡‡‡‡‡‡‡‡‡‡‡
   .4          ‡‡‡‡‡‡‡‡‡‡‡‡‡‡‡‡‡‡‡‡‡‡‡‡‡‡‡‡‡‡‡
   .5    1012.0 ‡‡‡‡‡‡‡‡‡‡‡‡‡‡‡‡‡‡‡‡‡‡‡‡‡‡‡‡‡‡
   .6          ‡‡‡‡‡‡‡‡‡‡‡‡‡‡‡‡‡‡‡‡‡‡‡‡‡‡‡
   .7          ‡‡‡‡‡‡‡‡‡‡‡‡‡‡‡‡‡‡‡‡‡‡‡‡‡‡
   .8          ‡‡‡‡‡‡‡‡‡‡‡‡‡‡‡‡‡‡‡‡‡‡‡‡
   .9          ‡‡‡‡‡‡‡‡‡‡‡‡‡‡‡‡‡‡‡‡‡‡
  1.0   1013.5 ‡‡‡‡‡‡‡‡‡‡‡‡‡‡‡‡‡‡‡‡‡
  1.1          ‡‡‡‡‡‡‡‡‡‡‡‡‡‡‡‡‡‡‡
  1.2          ‡‡‡‡‡‡‡‡‡‡‡‡‡‡‡‡‡
  1.3          ‡‡‡‡‡‡‡‡‡‡‡‡‡‡‡
  1.4          ‡‡‡‡‡‡‡‡‡‡‡‡‡
  1.5   1015.0 ‡‡‡‡‡‡‡‡‡‡‡‡
  1.6          ‡‡‡‡‡‡‡‡‡‡
  1.7          ‡‡‡‡‡‡‡‡‡
  1.8          ‡‡‡‡‡‡‡‡
  1.9          ‡‡‡‡‡‡
  2.0   1016.5 ‡‡‡‡‡
  2.1          ‡‡‡‡
  2.2          ‡‡‡
  2.3          ‡‡
  2.4          ‡‡
  2.5   1018.0 ‡
  2.6          ‡
  2.7          ‡
  2.8          ‡
  2.9          ‡
  3.0   1019.5 ‡
```

Figure 8-3. Standard Normal Distribution.

would like to guarantee that the containers have a full 1000 ml. The mean is set high enough so that virtually no container will have less that 1000 ml of fill.

The Standard Normal Distribution

In an earlier chapter we used Z-values to characterize the standing of an observation in a group. The standard score, or Z-value, tells us that an observation stands at that number of standard deviations above or below the mean for the group. We said then that Z-scores from -1.0 to $+1.0$ were common, while scores of less than -2.0 or more than $+2.0$ (more than two standard deviations from the mean) placed an observation toward the extremes of the group. Conversion to Z-scores makes the mean of any set of values zero and the standard deviation 1.0.

The normal distribution shape is defined in terms of Z. The ordinate at any point, compared to the height of the maximum ordinate at the mean ($Z = 0$) is

$$\frac{Y}{Y_{max}} = e^{-Z^2/2},$$

where e is the constant 2.71828, the base of natural logarithms. It is this negative exponential form where the argument of Z is squared that gives the curve of error its sinuous form in which the ordinates decrease rapidly up to Z squared $= 1.0$ and then decrease ever more slowly after that. De Moivre clearly gave this formulation of the normal law of error. His work was not immediately recognized by contemporaries, and other mathematicians continued to entertain other shapes as models for a law of error. For example, Thomas Simpson (1710–1761) used both a rectangular (equal probabilities of different size errors) and an equilateral triangular (probabilities inversely proportional to size of error) distribution as shapes for the law of error.

De Moivre derived the normal curve as the limit of the binomial distribution with $P = \frac{1}{2}$ as n indefinitely increases. This will be recognized as the principle underlying the probability board apparatus that we used in our example.

The normal distribution is a fixed form in relation to standard deviation units. The Z-scale is shown under the milliliter scale on our distribution. Each value on the Z-scale is

$$Z = \frac{X - \mu}{\sigma}.$$

We have constructed the distribution in Figure 8-3 so that there are 1000 asterisks. So, to find the probability that an observation equals or exceeds a given Z-score we need only count the symbols. Ten symbols equals probability 0.01, or 1 percent. Table 8-1 gives some probabilities in the right tail of the standard normal distribution. What is the probability that a container will

Table 8-1. Upper-Tail
Probabilities for the
Standard Normal
Distribution.

Z-score (z)	Probability $P(Z > z)$
0.67	0.2514
1.00	0.1517
1.28	0.1003
1.50	0.0668
1.65	0.0495
1.96	0.0250
2.00	0.0228
2.32	0.0102
2.58	0.0049
3.00	0.0013

be filled to 1016.5 ml or more? If we count the symbols at 1016.5 or greater, we find 22 out of 1000, or a little over 2 percent. The table above says that the probability in the tail of the standard normal distribution beyond 2.0 (the Z-value for 1016.5) is 0.0228. The probability of an observation at or above $Z = 3.0$ is very small (0.0013). This same value applies for the probability of an observation at or below -3.0, the left tail of the distribution.

The Z-values can be used to adjust risk levels. For example, if we are willing to run a 1 in 100 risk of filling a container with less than 1000 ml we could set the machine for a mean fill 2.32 standard deviations above 1000 ml, or at

$$1000 + 2.32(3.0) = 1007 \text{ ml.}$$

Many people have their first encounter with the normal distribution through grading practices in secondary schools. The following item appeared in *Newsweek Magazine*.

Next to the handwriting on a doctor's prescription, the least comprehensible language to a layman is educatorese, a form of gobbledygook that turns teeter-totters into recreational equipment and children of the same age into a peer group. In educatorese, Arcadia High School in the Scottsdale suburb of Phoenix, Arizona tried mightily last week to explain how 144 students made the honor roll. A mimeographed announcement posted on bulletin boards and sent to newspapers said:

'A student whose average exhibited performance in all credited subjects in relation to the performance of all other students falls at the level which places him (or her) on the normal curve of probability at a point falling on the plus side of the mean and between the second and third standard deviations will have made HONOR ROLL FIRST CLASS.'

Translation: "The top 2 percent," said Dr. Richard E. Bullington, principal of

Table 8-2. Operating Ratios (%) of Franklin and
Control Group Banks.

| | | Control Banks | | |
Year	Franklin	Mean	Standard deviation	Z-Score for Franklin
1969	84.06	78.27	5.40	1.07
1970	86.14	79.54	5.30	1.25
1971	91.19	81.04	5.80	1.75
1972	93.77	82.08	6.10	1.92
1973	95.60	84.90	5.16	2.07

the new $2 million plant in the prosperous Camelback Mountain area. Dr. Bullington ought to know. "I wrote it," he conceded in plain English and without apology. [4]

In another example high values rate attention but not commendation. The data are collected by the Comptroller of the Currency in connection with regulation of banks. The failure of the Franklin National Bank of New York on August 8, 1974 was the largest bank failure in U.S. history up to that time. The ratio of operating expense to operating income for the Franklin Bank, compared to a control group of 50 large banks, is given in Table 8-2 for the years preceding the failure. [5]

One assumes the control group banks are average to good performers. They form the criterion. The Z-scores tell us that Franklin Bank's operating ratio is more and more unusual for a well-operated bank as the 1970s progress. Indeed, if the control bank ratios are normally distributed, Franklin is in the extreme 5 percent by 1971. A monitoring procedure that singled out such extremes for further investigation might have been helpful in forestalling the failure.

Probability Distribution of the Sample Mean

As useful as the normal distribution is in analyzing individual measures, it has its widest use in analyzing means, or averages, of samples. Let us return to the container-filling example and ask how the operation of the machine might be monitored as it fills lot after lot of containers.

The answer is provided by sampling the lots for mean container fill and maintaining a quality control chart for the sample mean. Quality control charts were the special contribution of Walter A. Shewhart (1891–1967), who pioneered their application while with the Bell Telephone System Laboratories in the 1930s. The control chart is based on fundamentals concerning means of random samples.

The theory behind the control chart is that the probability distribution of the sample mean of a sample of size n from a normally distributed population with mean $= \mu$ and standard deviation $= \sigma$ is normal with a mean equal to the population mean and standard deviation equal to the population standard deviation divided by the square root of the sample size. This last very important formula appears to have been first stated by Gauss in 1809.

Figure 8-4 shows the probability distribution of the sample mean for samples of nine fills from the filling machine. We said the population of fills, when the machine is operating as intended, has a mean of 1010.5 and a standard deviation of 3.0 ml.

The probability distribution of the mean of a sample of $n = 9$ from the fills population has:

(1) a mean of 1010.5 (the population mean); and has
(2) a standard deviation $= 3.0/\sqrt{9} = 1.0$ ml; and
(3) is normal if the population of fills is normal.

The standard deviation of the probability distribution of the sample mean is also called the *standard error of the mean*, which we will abbreviate as $SE(\bar{x})$.

The distribution of Figure 8-4 is more highly concentrated around the population mean than the single-fill distribution in Figure 8-3. If we were talking about samples of size $n = 25$ the distribution would be tighter, and it would be even more concentrated for means of samples of size 100 because

$$\text{for } n = 25: \quad SE(\bar{x}) = 3.0/\sqrt{25} = 0.6,$$

$$\text{for } n = 100: \quad SE(\bar{x}) = 3.0/\sqrt{100} = 0.3.$$

This expresses what we would expect intuitively—that larger samples produce

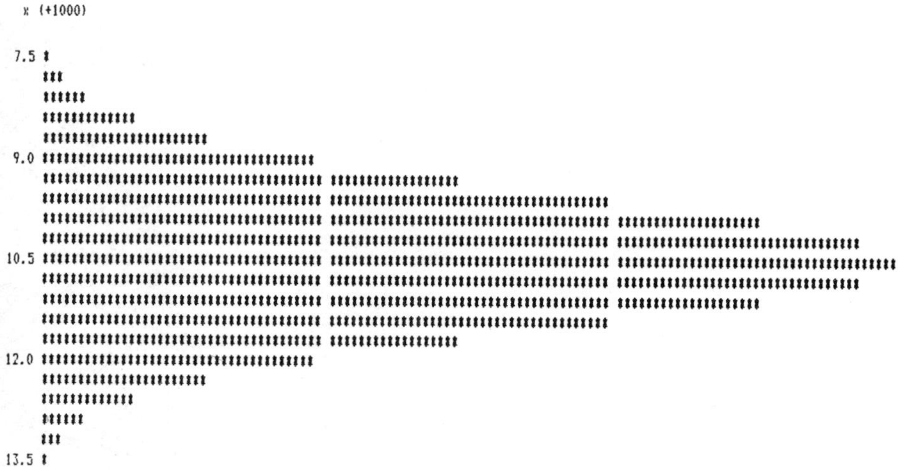

Figure 8-4. Probability Distribution of Sample Mean for $n = 9$ from Fills Population.

means that are on the average closer to the population mean than smaller samples.

Process Control Chart

The theory underlying the operation of a control chart for process control is shown in Figure 8-5. From each successive large lot of container fills a sample (in our example, $n = 9$) of fills are measured, and the sample mean deter-

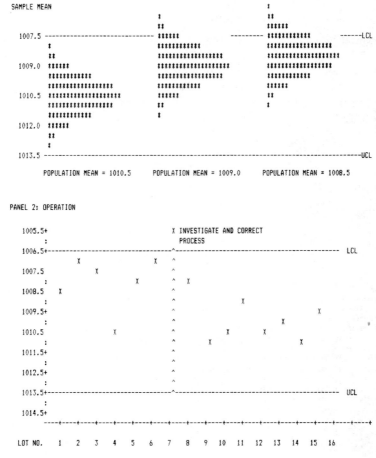

Figure 8-5. Control Charts.

mined. The sample mean is compared to the probability distribution of the sample mean that prevails when the machine is operating as intended. This is the distribution in Figure 8-4. It is shown again at the left in Figure 8-5.

At the Z-scores of plus and minus 3.0 on this distribution you see lines drawn for a *Lower Control Limit* (LCL) and an *Upper Control Limit* (UCL). Should a sample mean fall *below* the lower control limit or *above* the upper control limit, the process is declared *out-of-control*. Such sample means occur only very rarely when the process is operating as intended. When this event happens, the presumption is that something has happened to cause the process to depart from the long-run mean of 1010.5 ml of fill. The process would be halted and a search begun for the cause of the difficulty.

The two additional distributions in Figure 8-5 show the theory for the situation where a change in process mean has occurred. The process mean in the second distribution is 1009.0, and in the third the process mean is 1008.5 ml. fill. The chance of a sample mean ($n = 9$) falling below the established lower control limit when the process mean is 1009.0 can be found from the Z-score for that occurrence.

$$Z = \frac{\text{LCL} - \mu}{\text{SE}(\bar{x})} = \frac{1007.5 - 1009.0}{1.0} = -1.5.$$

From Table 8-1 we see that the normal "tail" probability for a Z-score of 1.5 is 0.0668. We are not very likely to catch this shift in mean fill.

The third distribution shows the situation for a shift in the process mean to 1008.5. The Z-score for the lower control limit is now

$$Z = \frac{\text{LCL} - \mu}{\text{SE}(\bar{x})} = \frac{1007.5 - 1008.5}{1.0} = -1.0,$$

and the "tail-area" probability is 0.1517. While the probability of detecting the shift on *one* sample of $n = 9$ is still not large, remember that samples will continue to be taken. If the shift to this level persists, the chances of its remaining *undetected after five samples of nine fills each have been checked is*

$$(1 - 0.1517)^5 = 0.4392.$$

Of course if a shift is more extreme we stand to catch it sooner.

An ongoing control chart looks like the second panel of Figure 8-5. The control limits (LCL and UCL) have been prefigured and drawn. Means of successive samples are plotted, and an out-of-control condition is signaled whenever the sample mean falls beyond the control limits.

Process quality control was a primary force behind the emergence of Japan as an industrial leader in the 1970s. In rebuilding Japan's industrial capacities, leaders recognized the importance of quality control. W. Edwards Deming, an American statistician, undertook systematically to teach quality control principles and techniques to the work force of the entire nation. Quality control and quality circles transformed Japanese products.

The Central Limit Theorem

In our quality control example we supposed the distribution of original measures to be normal. When this is so it follows that the distribution of sample means is normal regardless of sample size. Mathematical statisticians, beginning at least with Laplace in 1810 have been able to show that even when the distribution of single measures is not normal, the probability distribution of the sample mean will approximate normality more and more closely as the sample size, n, increases. The sample size required before normality can be assured is generally taken to be of the order of 25 or so.

The assurance provided by the central limit theorem greatly extends the applicability of the normal distribution in problems involving sample means. While it is still true that for small samples one must be concerned whether single measures have a normal distribution, if the sample size exceeds 25, then normality of the probability distribution of the sample mean is assured.

References

1. E. S. Pearson, "Some Notes on the Early History of Biometry and Statistics," *Biometrika*, **52** (1965), 3–18.
2. Helen M. Walker, *Elementary Statistical Methods*, New York, Holt, 1943, p. 166.
3. W. J. Youden, *Experimentation and Measurement*, National Bureau of Standards Special Publication 672, 1955.
4. *Newsweek Magazine*, November 30, 1959.
5. Joseph F. Sinkey, Jr., *Problem and Failed Institutions in the Commercial Banking Industry*, JAI Press, Greenwich, CT, 1979.

CHAPTER 9

Political Arithmetic

The word *statistics*, or state arithmetic, was coined by Gottfried Achenwall (1719–1772), in 1749. Achenwall was a professor at the universities of Gottingen and Marborough in Germany. By state arithmetic is meant the carrying on of those counting and calculating activities necessary to the operation of a modern nation-state. M. J. Moroney explains to us that rulers

> ... must know just how far they may go in picking the pockets of their subjects. A king going to war wishes to know what reserves of manpower and money he can call on. How many men need be put in the field to defeat the enemy? How many guns and shirts, how much food, will they need? How much will all this cost? Have the citizens the necessary money to pay for the king's war? Taxation and military service were the earliest fields for the use of statistics. [1]

Beginnings in England

John Sinclair (1754–1835) was the first English writer to use the term *statistics*. He had completed a comprehensive survey of Scotland in 1791 based on material from parish ministers. In the preface Sinclair observed that while *statistics* in Germany were concerned with the political strength of the state, he was more interested in

> ... the quantum of happiness enjoyed by the inhabitants and the means of its improvement; yet as I thought a new word (statistics) might attract more public attention, I resolved to adopt it. [2]

The beginnings of political arithmetic in the Anglo-American world are usually traced to John Graunt and William Petty. In 1661–1662 in England, Graunt (1620–1674) published his *Observations on London Bills of Mortality*.

These were commentaries on birth and death rates, cases of plague, and causes
of death based on records of vital statistics which were required to be maintained by the churches. William Petty (1623–1687) in *Essays in Political
Arithmetic*, published over the period 1683–1690, used this sort of information as a starting point in examining the growth of population, commerce
and manufactures in London as compared to Paris and Rome. Graunt constructed the first crude table of life expectancy, which was improved upon
by Edmund Halley (1656–1742) in 1690. John Arbuthnot (1667–1735), an
English mathematician, physician, and satirist, supported the emerging disci-
pline of political arithmetic around 1710 with these words:

> all the visible works of God Almighty are made in number, weight, and measure;
> therefore to consider them we ought to understand arithmetic, geometry, and
> statics. [3]

The Germans were not to be outdone in seeing the hand of God in things
statistical. To John Peter Sussmilch (1707–1767), who wrote a 1201-page
treatise on population, including 207 tables, God was the

> infinite and exact Arithmaticus . . . who has for all things in their temporal state
> their score, weight, and proportion. [4]

America to 1790

Arbuthnot seemed to despair that the Americans could measure up, however.
He said that nations that lack arithmetic are

> . . .altogether barbarous, as some Americans, who can hardly reckon above
> twenty. [5]

As in England, early records on vital statistics were kept by the churches in
America. Early studies were done by clergy who maintained these records.
Notable among these were Ezra Stiles, pastor of a Congregational church in
Newport, Rhode island, and the Reverend Jeremy Belknap of New Hampshire. In 1787 Stiles published an analysis of the vital statistics of the New
Haven, Connecticut church community from 1758 to 1787—births, deaths,
baptisms, marriages. Deaths were grouped by 5-year age class and by causes.
Belknap published a *History of New Hampshire* between 1784 and 1791 that
contained commercial and governmental as well as population statistics for all
towns in the state. Using his and others' estimates for early years, Belknap
estimated that the state's population had doubled in the 19 years ending in
1790, the year of the first Federal census—despite an estimated 1400 deaths
from the Revolutionary War.

The Founding Fathers were concerned with political arithmetic, particularly the population question. Benjamin Franklin wrote about population increase in *Poor Richard's Almanac* and elsewhere. In 1766 Franklin visited the
University of Gottingen, where Gottfried Achenwall quizzed him to obtain

information about American cities, commerce, and population. In 1780 John Adams sought aid from Holland for the American Revolution, and was asked about the population base for the American undertaking. His reply was

> It has been found by calculation, that American has doubled her numbers, even by natural generation alone ... about once in eighteen years ... There are nearly twenty thousand fighting men added to the numbers in America every year. Is this the case of our enemy, Great Britain? Which then can maintain the war the longest? [6]

In 1786 Thomas Jefferson wrote that

> The present population of the inhabited parts of the United States is about ten to the square mile; and experience has shown us, that wherever we reach that, the inhabitants become uneasy, as too much compressed, and so go off in great numbers to search for vacant country. Within forty years their whole territory will be peopled at that rate. We may fix that, then, as the term beyond which the people of those (western) states will not be restricted within their present limits. [7]

Seventeen years later President Jefferson doubled the area of the continental United States with the Louisiana Purchase.

Alexander Hamilton, in advocating the assumption of existing state debt by the federal government in 1790 and 1791, drew on the actuarial work of Richard Price and Edmund Halley. This is the same Dr. Price who read Thomas Bayes' essay on a theorem in probabilities to the British Royal Society and Dr. Halley, British astronomer and mathematician who discovered Halley's comet.

The United States, 1790–1900 [8]

Authority for conducting a census of population in the United States is contained in Article I of the Constitution. The article directs that a census be conducted every 10 years for the purpose of apportioning seats in the House of Representatives. The first Census of Population was duly conducted in 1790.

The major figures in statistics in the United States in the nineteenth century played a key role in the development of the census operations from a sporadic every tenth year political activity to a full-time professional organization. The first of these was Lemuel Shattuck (1793–1859). Shattuck's interest in statistics began when he encountered imperfect records of births and deaths while preparing *A History of the Town of Concord* (Massachusetts). As a member of the Common Council of Boston in 1837–1841 he urged the city fathers to authorize a census because the federal census of 1840 contained serious errors. He was appointed to direct a census of Boston in 1845, and was later called to Washington to assist in plans for the 1850 federal census. He drafted the questionnaires for the 1850 census and wrote instructions for the enumerators. Scholars regard the 1850 census as the first one of real value. In 1839 in

Boston, Shattuck had been instrumental in the founding of the American Statistical Association. Shattuck maintained correspondence with European scholars in statistics, including Adolphe Quetelet.

Edward Jarvis (1803–1884) and James D. B. DeBow (1820–1867) made contributions to the 1850 census as well. Jarvis was appointed to edit the vital statistics of the 1860 census and was consulted extensively in preparations for the 1870 census. He was a Massachusetts physician who specialized in the treatment of mental patients. He was President of the American Statistical Association from 1852 to 1882.

DeBow was the publisher of *DeBow's Review* from 1846 to 1880, which served to "collect, combine, and digest" important statistics of the South and West. He was placed in charge of census planning by President Franklin Pierce (1852). In this capacity he edited the census reports of 1850 and supervised a compendium entitled *Statistical View of the United States*, which was published in 150,000 copies.

Francis A. Walker (1840–1897) was appointed to head the 1870 census after reorganizing the Bureau of Statistics in the federal government. He was reappointed superintendent of the 1880 census, in which for the first time the superintendent had the power to appoint enumerators. Walker held professorships in economics and statistics at Yale and at Johns Hopkins. The 1880 census was highly regarded and secured his reputation as a statistician. From 1881 he held the Presidency of Massachusetts Institute of Technology, and published extensively in economics. He was President of the American Economic Association, from its founding in 1885, to 1892, and President of the American Statistical Association from 1883 to 1897. He advocated a permanent bureau of the census (achieved in 1902), pioneered in graphic presentation, including color, urged the recognition of statistics in collegiate curricula, and contributed to public awareness of the value of statistical information through articles in magazines such as the *Atlantic Monthly*.

This review of some leading contributors to federal census activities takes us into the 1890s. The membership of a special committee of the American Economic Association which published a 516-page volume in 1899 entitled *The Federal Census: Critical Essays by Members of the American Economic Association* includes some others who bear mentioning. Richmond Mayo-Smith (1854–1901), was founder of the Graduate School of Political Science at Columbia University in 1880 and author of important early texts in statistics, *Statistics and Sociology* (1895) and *Statistics and Economics* (1899). Carroll D. Wright (1840–1909) was the first United States Commissioner of Labor Statistics.

Statistics and Social Issues

Two persons on opposite sides of the Atlantic toward the end of the nineteenth century bring attention to the role of statistics in social and political issues.

They were an American, Carroll D. Wright (1840–1909) and an Englishman, Charles Booth (1840–1916).

By the last half of the nineteenth century the industrial revolution had brought about conditions in the factory cities and towns that concerned and alarmed many people. Working conditions, especially for women and children, poor housing, drunkenness, and destitution seemed to be the by-product of economic progress. The Poor Law of 1834 in England seemed to embody the idea of Thomas R. Malthus (1766–1834) that to help the destitute would only cause their numbers to multiply and thereby increase and not decrease misery. Malthus had argued that population tends to increase geometrically (as 1, 2, 4, 8, 16, etc.) while the means of subsistence increases only arithmetically (as 1, 2, 3, 4, 5, etc.). The ultimate check on population growth has to come in the form of war, pestilence, and famine. In both England and the United States the prevailing Protestant work ethic, buttressed by survival-of-the-fittest ideas taken from Darwinian evolutionary theory, caused many to ignore what Thomas Carlyle in England called "the condition of the people question" and what reformers in the United States called "the labor problem."

Pressures from the trade union movement and reformist groups in Massachusetts caused the legislature to establish in 1869 a Bureau of Statistics of Labor whose charge was

> to collect, assort, systematize, and present ... to the legislature ... statistical details relating to all departments of labor ... especially to the commercial, educational, and sanitary conditions of the laboring classes, and to the permanent prosperity of ... productive industry. [9]

In 1873, after a disastrous experience with the first chief, Carroll D. Wright was appointed head of the Bureau. He had been a colonel in the Civil War, a teacher, lawyer, and a state senator. Wright sought the advice of Francis A. Walker in matters of staffing and operating philosophy. As a result, statistical expertise was brought to bear on studies of wages and the cost of living. By 1884, fourteen states had followed the lead of Massachusetts in establishing bureaus of labor statistics.

At the 1881 convention of the Federated and Organized Trades and Labor Unions (later the American Federation of Labor), Samuel Gompers had urged the establishment of a national bureau of labor statistics. In 1884 Congress passed an act creating the Bureau of Labor in the Department of the Interior, and in 1885 President Chester A. Arthur appointed Carroll D. Wright to the post of commissioner. Wright was well known as a statistician from his work in Massachusetts and with the census. He had written a report on the factory system as part of the publications of the 1880 census. Wright ultimately served five 4-year terms as commissioner. Early on, he had expressed his view of his responsibility in these words:

> A bureau cannot solve social or industrial problems ... but its work must be classed among educational efforts, and by *judicious investigations and fearless*

publication of the results thereof, it may and should enable the people to more clearly and more fully comprehend many of the problems which now vex them. [10]

The phrase in italics (added) became in time a by-word in the Bureau of Labor Statistics. Wright was called on to supervise the Census of Manufactures of 1890. Wright is generally credited with the famous phrase "figures won't lie, but liars will figure."

Charles Booth was a Liverpool businessman who set out to refute what he regarded as cheaply sensational reports on poverty that were being circulated by socialists. He ended up producing from 1877 to 1903 a 17-volume study entitled *Life and Labour of the People of London*. His initial surveys were conducted with the help of School Board truant officers, who knew their districts in great detail. He had written to his assistant, Beatrice Potter (who later married the Fabian Socialist, Sidney Webb).

> What I want is to see a large statistical frame-work ... built to receive accumulations of fact out of which at last is evolved the theory and the law and the basis of more intelligent action. [11]

Booth set out to establish what he called the *poverty line*, and eventually found "one-third of the population sinking into want." More importantly, his work refuted the idea that the poor would breed up to the limits of subsistence. He found that increments in income were accompanied by more successful control of the birth rate.

Up to the last decade of the 1800s, there was little contact between the academic side of statistics, which was almost entirely mathematics, and the attempts to gain an understanding of social problems through statistics, which were the province of collectors of data who were generally untrained in the mathematical side of statistics. The field of applied social statistics was not represented in universities.

Samuel H. Stouffer, a well-known sociologist, wrote in 1938 of Walker and Wright.

> In the 1880s research in economics, sociology, political science, and education had hardly begun ... The social science research in Francis A. Walker's census of 1880, and in publications of the 1890 census completed under the supervision of Carroll D. Wright, probably was more significant, both in quantity and quality, than all the university research of the period put together. [12]

This state-of-affairs was about to draw to a close. While Mayo-Smith, mentioned earlier, was working on his texts, Walter Willcox (1861–1964) inaugurated a course at Cornell University in what was later called "social statistics." During his 40 years at Cornell, Willcox served as advisor to the Bureau of the Census, the New York State Board of Health, and a variety of international bodies concerned with population, labor, and health matters. [13]

In England as well, the time was drawing closer when the mathematical methods and the data collections of social statistics would be joined under the aegis of departments of applied statistics led by people who were making contributions to the practice as well as to the theory of statistics. First among these was Karl Pearson, who drew extensively from the materials in Booth's studies of London to illustrate his lectures.

1900–1950

By 1902, when the Census Office was established as a permanent agency, censuses of manufactures, mineral industries, agriculture, and governments were being conducted every 10 years in addition to the population census. Manufactures had been initiated in 1810, mineral industries and agriculture in 1840, and governments in 1850.

In 1910, under the leadership of Walter Laidlaw, census tracts were outlined in New York City. Census tracts are small areal units for analysis of population and (later) housing data. They form a basis for setting up local administrative or business areas for the analysis of individual firm or agency data and their comparison to the census population and housing data. The earliest applications were connected with the administration of social services in the cities.

In 1913 the Bureau of the Census was located in the Department of Commerce, its present home.

The economic and manpower planning associated with participation in World War I hastened the development of statistical programs. The Bureau of Labor Statistics began its index number programs under a requirement from the shipbuilding industry to develop a cost-of-living index for urban wage earners, although regular publication of what is now called the Consumer Price index did not begin until 1921. Economist–statisticians who contributed most significantly to this field were Wesley C. Mitchell (1874–1948) of Columbia and Irving Fisher (1867–1947) of Yale University, the author of *The Making of Index Numbers* (1922).

In the 1920s Arthur Burns and Wesley C. Mitchell conducted extensive studies of U.S. business cycles for the National Bureau of Economic Research, a privately endowed research organization then housed in New York City. These and successor studies led to the development of an index of leading business indicators for anticipating cyclical turns in the economy. The bureau began its research into national income measurement in 1921.

During the 1920s the Bureau of Agricultural Economics in the Department of Agriculture conducted a number of studies of farm costs and prices in response to the agricultural difficulties of that time. Henry A. Wallace (later Vice President under Franklin D. Roosevelt) was an editor of *Wallace's Farmer* in Des Moines, Iowa. His father, Henry C. Wallace, was Secretary of Agriculture from 1921 until 1924. Henry A. became interested in the statistical methods, largely multiple regression and correlation, used in the studies, and in machine computations. In 1924 he organized a series of Saturday seminars at Iowa State University to explain these new techniques and technology to statistics faculty and graduate students under G. W. Snedecor. [14]

When Henry A. Wallace became Secretary of Agriculture in 1933 he appointed Mordecai Ezekiel (1899–1974) Economic Advisor to the Secretary. Ezekiel had been in the Bureau of Agricultural Economics during the 1920s and was a leader in developing the methods used for them. His *Methods of Correlation and Regression Analysis*, first published in 1931, was an influencial text (later revised with Karl A. Fox in 1959) for a generation. [15] Wallace supported a strong statistical tradition in the Department of Agriculture and close ties with the Statistics Department at Iowa State University.

The first censuses of construction, retail, and wholesale trades were inaugurated by the Bureau of the Census in 1930, and in 1933 the first census of selected service industries was conducted.

In response to a Senate Resolution in 1932, the task of developing national income measures was assigned to the Department of Commerce. Simon Kuznets, of the University of Pennsylvania and the National Bureau of Economic Research, assumed the directorship of the work in 1933, and the first report on national income (from 1929 to 1932) was submitted to the Senate on January 4, 1934 [16]. Kuznets continued to contribute to work in national income measurement as an advisor to the Commerce Department. In 1971 he received the Nobel Prize for his work in the measurement of economic growth.

State-by-state income payments were first published by the Department of Commerce in 1939 for the years 1929–1937. The measurement of product flows became important in World War II in connection with war mobilization analysis, and led to the first publication of gross national product figures by the Department of Commerce in 1942. [17]

The first Census of Housing was conducted in 1940 along with the Census of Population. Beginning in 1940 the census tract programs were developed and expanded—opening up many avenues for the local use of census and related data.

The year 1950 saw the initiation of sampling as an integral part of the census as certain detailed housing characteristics were collected from only one-in-four repondents. Over the period from 1933 to 1950 large-scale sampling methods were developed by leading survey statisticians in the Bureau of the Census. First among these was Morris Hansen.

Roughly the same period also saw the development of input–output analysis by Wassily Leontief at Harvard University. It was not until the late 1950s and 1960s that input–output analysis became commonplace in government programs, however. Leontief received the Nobel Prize in economics in 1973 for his work in input–output analysis.

Since 1950

While computers were used in tabulating the 1950 censuses of population and housing, it was not until 1960 that the entire census was tabulated by computer. The development of FOSDIC (Film Optical Sensing Device for Input to Computer) during the 1950s was an important prerequisite. In the 1960 censuses mail-back self-enumeration forms were introduced in urban areas. Self-enumeration was extended in the 1970 censuses and in 1980 around 90 percent of households received mail-back forms—thus greatly reducing the armies of enumerators previously used. During the 1960s computerized geographic base files were developed for analysis and mapping of urban area data. The first Census of Transportation was conducted in 1963, and in 1973 the Census Bureau began its Annual Housing Survey. [18]

In 1959 responsibility for labor force statistics was shifted from the Census Bureau to the Bureau of Labor Statistics. A major vehicle for producing estimates of employment and unemployment is the Current Population Survey, which was established by the Census Bureau in 1947 and continued after 1959 under contract to BLS. The sample consists of some 60,000 households across the nation. [19]

Labor market data are also collected through a BLS program conducted in cooperation with state employment security agencies called the *Current Employment Statistics* (CES) program. Data on employment, hours, and earnings are collected from a national sample of almost 200,000 establishments. While the basic survey goes back to 1919, a major improvement in the ability to estimate national, regional, and statewide totals from the data was made in 1949 when BLS was able to tie changes from the CES data to *benchmarks* provided by employer tax files from the *Unemployment Insurance* (UI) system. In the 1960s BLS assisted the state employment security agencies in implementing improved methods for sampling of establishments. [20]

In the post-World War II period the Bureau of Labor Statistics continued to improve its price index procedures. Surveys of consumer expenditure patterns were conduced in 1950, 1960–1961, 1972–1973, and 1984–1985 to update the market basket utilized for weights in the Consumers Price Index (CPI). In 1978 a number of refinements were introduced in selecting outlets for periodic pricing of items. [21] In this same revision of procedures, BLS instituted the all urban consumers' price index (CPI-U) to augment the older index for wage and clerical workers (CPI-W). An important recent change was to treat homeowner housing services in terms of rental equivalents in order to avoid distortions in the price indexes previously brought about by changes in the *investment* values of homes.

References

1. M. J. Moroney, *Facts From Figures*, 3rd edn., Penguin, Baltimore, MD, 1956, p. 1.
2. G. Udney Yule and M. G. Kendall, *An Introduction to the Theory of Statistics*, Hafner, New York, 1950, p. xvii.
3. Quoted in Harry Hopkins, *The Numbers Game*, Little, Brown, Boston, MA, 1973, p. 21.
4. Quoted in James H. Cassedy, *Demography in Early America*, Harvard University Press, Cambridge, MA, 1969, p. 105.

5. Quoted in Hopkins, op. cit., p. 23.
6. Quoted in Cassedy, op. cit., p. 205.
7. Quoted in Cassedy, op cit., p. 229.
8. A major source for this section is Paul J. Fitzpatrick, "Leading American Statisticians in the Nineteenth Century," *Journal of the American Statistical Association*, **52**, No. 279 (September 1957), 301–321.
9. Ewan Clague, *The Bureau of Labor Statistics*, Praeger, New York, 1968, p. 5.
10 Ibid., p. 11.
11 Quoted in Hopkins, op. cit., p. 29.
12 Quoted in Meyer H. Fishbein, "The Census of Manufactures, 1810–1890," *National Archives Accessions*, No. 57 (June 1963), 20.
13. William R. Leonard, "Walter F. Willcox: Statist," *The American Statistician*, **15** (February 1961), 16–19.
14. Jay L. Lush, "Early Statistics at Iowa State University," in *Statistical Papers in Honor of George W. Snedecor* (T. A. Bancroft, ed.) Iowa State University Press, Ames, IA, 1972, pp. 211–226.
15. Oris V. Wells, "Mordecai J. B. Ezekiel, 1899–1974," *The American Statistician*, **29**, No. 2 (May 1975), 106.
16. Carol S. Carson, "The History of the United States National Income and Product Accounts: The Development of an Analytical Tool," *Review of Income and Wealth*, **Ser. 21** (June 1975), 153–181.
17. George Jaszi, "Bureau of Economic Analysis," in *Encyclopedia of Economics*, (Douglas Greenwald, ed.) McGraw-Hill, New York, 1982.
18. *U.S. Bureau of the Census*, "Factfinder for the Nation," CFF No. 4 (Rev.), May, 1979.
19. Janet L. Norwood, "Unemployment and Associated Measures," in *The Handbook of Economic and Financial Measures* (Frank J. Fabozzi and Harry I. Greenfield, eds.), Dow Jones-Irwin, Homewood, II, 1984, pp. 143–163.
20. Janet L. Norwood and John F. Early, "A Century of Methodological Progress at the U.S. Bureau of Labor Statistics," *Journal of the American Statistical Association*, **79**, No. 388 (December 1984), 748–761.
21. Ibid., p. 755.

CHAPTER 10

Regression and Correlation

Most people are acquainted with linear relationships. If a salesperson is paid $2000 a month plus a commission of 5 percent of sales, monthly income can be expressed as

$$Y = \$2000 + 0.05X,$$

where X is the sales for the month. If no sales occur, $X = 0$ and $Y = \$2000$; if sales are $20,000, then $Y = \$2000 + 0.05(\$20,000) = \$3000$. What we have here is a mathematical relationship expressing an agreement about compensation.

Statistical relationships are often linear. An early study in the United States involved the relationship between the age and heights of 24,500 Boston schoolboys. It was conducted for the Massachusetts Board of Health in 1877 by Henry P. Bowditch. [1] Bowditch was Superintendent of the Boston School Board. Figure 10-1 is derived from part of a table in Bowditch's study. Each age class contains about 1400 boys, and we see the distribution of heights at each age. The methods we will discuss were not available in Bowditch's time, but he came close to the ideas of regression and correlation, which were developed within a few years of his work.

In the salesperson income relationship, the salesperson receives an additional 5 cents in income for every additional dollar of sales. In a height–age relationship we would want to know the *average* addition to height for each additional year of age. This would be determined in some way from the data, and is a statistical average rather than a mathematical constant.

Another mathematical example comes from break-even analysis. Suppose a child sets up a lemonade stand, investing $5 in lumber for the stand. Mother agrees to supply lemonade and cups, and the child decides to charge $0.25 per

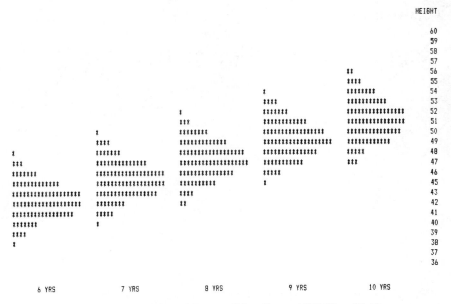

Figure 10-1. Heights of Boston Schoolboys, 1877 (Bowditch).

cup. The net income relation is

$$Y = -\$5 + \$0.25X,$$

where X is the number of cups sold. Income will be zero (the break-even point) at

$$X = \frac{\$5}{\$0.25} = 20 \text{ cups.}$$

A similar relationship, but this one statistical, was determined for the net earnings of major league baseball clubs in 1976. [2] The equation was

$$Y = -\$2,000,000 + \$2.08 \times \text{attendance.}$$

This was an average relationship for 26 baseball franchises. The *average* addition to income for each additional person attending a game during the season was \$2.08. The average costs unrelated to attendance were \$2 million per franchise. So, the *average* break-even attendance for a club in 1976 was

$$\text{Attendance} = \frac{\$2,000,000}{\$2.08} = 961,538 \text{ persons.}$$

For our specific example to illustrate the arithmetic of regression and correlation we turn to baseball as well. Table 10-1 gives the numbers of runs scored during the 1975 and 1976 seasons by American League clubs. We

Table 10-1. Regression for 1976 Runs Based on 1975 Runs for American League Baseball Clubs.

Team	Season Runs 1975 X	Season Runs 1976 Y	$(X - \bar{X})^2$	$(Y - \bar{Y})^2$	$(X - \bar{X})*(Y - \bar{Y})$	Yc	$(Y - Yc)$	$(Y - Yc)^2$
Yankees	681	730	82.51	7042.01	−762.24	639.55	90.45	8181.14
Orioles	682	619	65.34	733.51	218.92	640.27	−21.27	452.39
Red Sox	796	716	11218.34	4888.34	7405.34	722.26	−6.26	39.21
Brewers	688	615	4.34	966.17	64.76	644.58	−29.58	875.27
Indians	675	570	227.51	5788.67	1147.59	635.23	−65.23	4255.60
Tigers	570	609	14420.01	1375.17	4453.09	559.72	49.28	2428.91
A'S	758	686	4612.67	1593.34	2711.01	694.93	−8.93	79.76
Twins	724	743	1150.34	9392.84	3287.09	670.48	72.52	5259.56
White Sox	655	586	1230.84	3610.01	2107.92	620.85	−34.85	1214.55
Rangers	714	616	572.01	905.01	−719.49	663.28	−47.28	2235.86
Royals	710	713	396.67	4477.84	1332.76	660.41	52.59	2765.92
Angels	628	550	3854.34	9232.01	5965.17	601.43	−51.43	2645.17
Average	690	646	3152.91	4167.08	2267.66			2536.11

$\bar{X} = 690.0833$
$\bar{Y} = 646.0833$

$\text{Var}(X) = 3152.91$
$\text{Var}(Y) = 4167.08$
$\text{Cov}(XY) = 2267.66$

$b = 2267.66/3152.91 = 0.719228$
$a = 646.0833 - 0.719228(690.0833) = 149.75607$

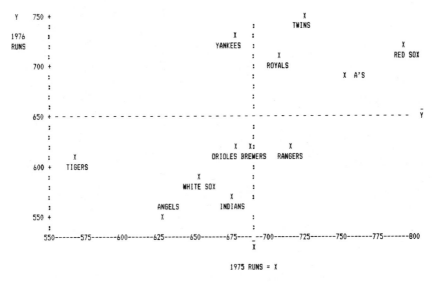

Figure 10-2. Total Runs by American League Baseball Teams, 1975 and 1976.

would expect some relationship in the sense that the more powerful teams (scoring more runs) in 1975 would tend to be the more powerful teams in 1976 as well. There should be some consistency of performance from year to year. But this will be far from perfect, as both disappointed and surprised managers can attest. In Figure 10-2 the data are plotted, and we see there a tendency for run production in 1976 to be associated with run production in 1975. In what follows we give numerical expression to the relationship, or association.

The Regression Line

One use that is made of relationships is to calculate, estimate, or predict the value of one of the variables that will occur when the other variable has a particular value. We may want to know the salesperson's income when sales are $50,000, or the child's net earnings when 50 cups of lemonade are sold. In both of these cases the relationship is exact and mathematical, and we need only substitute the appropriate X in the equation to find the corresponding Y.

When the relationship is statistical, as in the height–age and baseball franchise earnings–attendance relationships, we will have to find the line of relationship from the data. This is called the *regression line*, and we are dealing for the moment with straight-line, or linear regressions.

We want a line that in some sense is a *best fitting* line to the data points of Figure 10-2. The traditional criterion of best fit is the criterion of *least squares*.

This criterion says that we want the constants of the straight line

$$Yc = a + bX,$$

that makes the sum of squares of the actual points, Y, from the line, Yc, a minimum. It is an exercise in mathematics to derive equations for the coefficients a and b from the statement of the criterion. The criterion is to minimize S, where

$$S = \sum (Y - Yc)^2 = \sum [Y - a - bX]^2.$$

The criterion of least squared deviations and the derivation of a system of equations for finding the least squares line was first published by Adrien Marie Legendre (1752–1833) in a book called *New Methods for Determining the Orbits of Comets.* [3]

Application of the calculus will provide formulas for a and b that minimize the squared deviations. The formulas we show are equivalent and, we think, the most helpful for understanding regression and correlation. The *regression coefficient*, b, is found by dividing the *covariance* of X and Y by the variance of X. The covariance of X and Y is

$$\text{COV}(XY) = \frac{\sum (X - \bar{X})(Y - \bar{Y})}{n}.$$

This can be seen to be the average cross-product of deviations of the two variables from their respective means. It has some standing of its own as a measure of association. The covariance will be positive when values *above* the mean of Y are paired with values *above* the mean of X and negative deviations for Y are paired with negative deviations for X. From Figure 10-2 we can see that this is essentially the case for the baseball data. If Y tends to decrease as X increases, then the products of deviations will be on balance negative. If neither tendency is present, the signs of the cross-products will be mixed and the covariance will tend toward zero. Figure 10-3 shows such a situation.

We are already familiar with the standard deviation of a set of values. The variance of a set of values is just the square of the standard deviation. For a variable, X, the variance is

$$\text{Var}(X) = \frac{\sum (X - \bar{X})^2}{n}.$$

The regression coefficient can be expressed as

$$b = \frac{\text{COV}(XY)}{\text{VAR}(X)}.$$

Then, the coefficient, a, for the *least squares* regression line is obtained from

$$a = \bar{Y} - b\bar{X}.$$

In Table 10-1 showing the total runs for each American League team in 1975

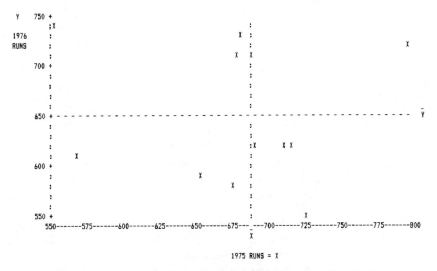

Figure 10-3. Baseball Data Rearranged to Show Zero Association.

(X) and 1976 (Y), we show the calculation of COV(XY) and VAR(X) following the formulas given. It is all very quickly done on a microcomputer spread-sheet program. We get the regression line (rounded) for estimating 1976 runs based on 1975 runs.

$$Yc = 150 + 0.72X.$$

The constant, 150, represents the Y-intercept in the slope-intercept equation for the straight line. The regression coefficient says that for every additional run in 1975, teams on the average scored an additional 0.72 runs in 1976. In Figure 10-4 the regression line is shown.

A useful way to orient one's thinking about a regression line is to think of it as originating at the point of intersection between the mean of Y and the mean of X. The least squares regression line always goes through this point. The regression can be stated, then, as

$$Yc = \bar{Y} + b(X - \bar{X}),$$

or, in our case,

$$Yc = 646 + 0.72(X - \bar{X}).$$

Now, the regression line says that if a team's runs were at the average in 1975, we would expect the team to have 646 runs in 1976. But, for every run above average in 1975 we would expect 1976 runs to increase by 0.72. Likewise, for every run below the 1975 average, 1976 runs are expected to decrease by 0.72.

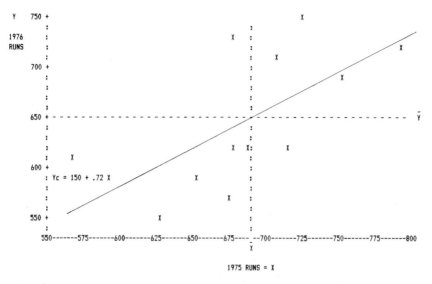

Figure 10-4. Total Runs by American League Baseball Teams, 1975 and 1976 with Regression Line.

Variation from Regression

As we have seen, the regression line is a line of average relationship. In a case of height–age it tells us the average height for persons of a given age (and sex). In that sense it is a sort of sliding average, or statistical norm. The measure of variation from this sliding average is very much like the standard deviation. What is different is that the regression line is the base for measuring deviations rather than the overall mean. The measure is called the *standard error of estimate*, and is defined as

$$\text{SE Est} = \sqrt{\frac{\sum(Y - Yc)^2}{n}}.$$

In Table 10-2 we show the calculation of the standard error of estimate for the baseball runs example. It is 50.4 runs. The standard deviation of the 1976 season runs for the 12 teams was 64.6 runs. The variation around the line of average relationship is somewhat smaller than the variation around the fixed average. This emphasizes what we knew from the graph—that there was a moderate relation between 1975 and 1976 team runs. We will make further use of the standard error of estimate later.

To the extent that the standard error is small we have a close relation between Y and X that will permit Y to be predicted or estimated closely from values of X. Should SE Est be virtually zero we are approaching the case of a mathematical rather than a statistical relationship.

Table 10-2. Unexplained, Explained, and Total Sums of Squared Deviations and Coefficient of Determination for Baseball Runs Data.

Team	Season Runs		Y_c	$(Y - Y_c)$	$(Y_c - \bar{Y})$	$(Y - \bar{Y})$	$(Y - Y_c)^2$	$(Y_c - \bar{Y})^2$	$(Y - \bar{Y})^2$
	1975 X	1976 Y							
Yankees	681	730	639.54	90.45	-6.53	83.92	8181.14	42.68	7042.01
Orioles	682	619	640.25	-21.27	-5.81	-27.08	452.39	34.80	733.51
Red Sox	796	716	722.24	-6.26	76.18	69.92	39.21	5803.12	4888.34
Brewers	688	615	644.57	-29.58	-1.50	-31.08	875.27	2.25	966.17
Indians	675	570	635.22	-65.23	-10.85	-76.08	4255.60	117.69	5788.67
Tigers	570	609	559.70	49.28	-86.37	-37.08	2428.91	7459.31	1375.17
A's	758	686	694.91	-8.93	48.85	39.92	79.76	2386.08	1593.34
Twins	724	743	670.46	72.52	24.39	96.92	5259.56	595.06	9392.84
White Sox	655	586	620.84	-34.85	-25.23	-60.08	1214.55	636.70	3610.01
Rangers	714	616	663.17	-47.28	17.20	-30.08	2235.86	295.89	905.01
Royals	710	713	660.39	52.59	14.32	66.92	2765.92	205.19	4477.84
Angels	628	550	601.42	-51.43	-44.65	-96.08	2645.17	1993.81	9232.01
Average	690.08	646.08				Sums	30433.35	19571.58	50004.92

$\bar{X} = 690.0833$
$\bar{Y} = 646.0833$

$Y_c = 149.75607 + 0.719228\,X$
Unexplained SSD = 30433.35
Explained SSD = 19571.58

Total SSD = 50004.92

Std Dev $(Y) = \sqrt{50004.92/12} = 64.55$
SE Est $= \sqrt{30433.35/12} = 50.36$

$r^2 = 19571.58/50004.92 = 0.39$

The *coefficient of determination* is a measure in regression that breaks down total variation in Y into two elements. These elements are variation *explained* by the regression line and variation *unexplained* by the regression. To introduce these concepts, consider the case of the Kansas City Royals in the following way.

(1) The average team runs in 1976 was 646. The Royals had 713 runs, 67 above the average.

$$\text{Total variation} = Y - \bar{Y} = 67.$$

(2) The Royals had 710 runs in 1975, from which the regression would predict $150 + 0.72(710) = 661$ runs in 1976. Actual runs in 1976 exceeded this by $713 - 661 = 52$.

$$\text{Unexplained variation} = Y - Yc = 52.$$

(3) By the same token, however, the regression has explained the difference between the overall average of 646 runs in 1976 and 661, the expected 1976 runs for a team with 710 runs in 1975.

$$\text{Explained variation} = Yc - \bar{Y} = 15.$$

For an aggregate measure involving these concepts we make use of the fact that these elements of variation add up in their squares, namely

$$\text{Unexplained sum of squared deviations} = \sum(Y - Yc)^2$$
$$+ \text{Explained sum of squared deviations} = \sum(Yc - \bar{Y})^2$$
$$= \text{Total sum of squared deviations} = \sum(Y - \bar{Y})^2.$$

The coefficient of determination, which goes by the symbol, r^2, is just the ratio of the explained sum of squared deviations to the total sum of squared deviations.

$$r^2 = \frac{\sum(Yc - \bar{Y})^2}{\sum(Y - \bar{Y})^2}.$$

In Table 10-2 this calculation is carried out and we find that the coefficient of determination is 0.39. Thirty-nine percent of the variation in 1976 runs is explained by the relation between 1976 runs and 1975 runs.

Correlation and Galton

In an earlier section we talked about expressing values of a variable in terms of Z-scores. The Z-score form standardizes variables so they have zero mean and unit (1.0) standard deviation. An individual Z-score indicates the standing of an observation as so many standard deviation units above or below the mean.

If one does regression with variables expressed in Z-scores, the regression coefficient will be

$$r = \frac{\sum Z(X)*Z(Y)}{n} = \frac{\sum (X - \bar{X})*(Y - \bar{Y})}{nS(X)*S(Y)}.$$

The final expression above is known as the Pearson product-moment formula for the correlation coefficient, r. It can be seen to be the covariance (average cross-product of deviations) divided by the product of the standard deviations of X and Y. Since least squares linear regressions go through the intersection of mean X and mean Y, the regression, in standardized form is

$$Zc(Y) = 0 + r * Z(X),$$

or just

$$Zc(Y) = r * Z(X).$$

In the baseball example

$$Zc(Y) = 0.623Z(X).$$

Keep in mind that 0.623 is the regression slope in standard deviation units. This means, for example, that if a team's runs were 1 standard deviation above the mean runs in 1975, we would expect 1976 runs to be 0.623 standard deviations above the mean. For a team at 2 standard deviations below the mean in 1975 the expected 1976 runs are $2(0.623) = 1.26$ standard deviations below the mean.

Here we have the historic meaning of regression. The relationship in standard deviation units emphasizes to us that the teams *regress* toward the mean. That is, teams above average in 1975 will tend to be above average in 1976, but not so far above average as they were in 1975. Teams below average in 1975 tend to move toward the average in 1976.

Francis Galton (1822–1911) receives credit for the inspiration that led to correlation and regression. Galton was a cousin of Charles Darwin. After his education at King's College in London and Trinity College in Cambridge, he began the study of medicine. Upon the death of his father he gave up medical studies and spent the next several years leading the life of an English gentleman. In 1850 he began a 2-year exploration in Southwest Africa. He was made a Fellow of the Royal Society in 1856 on the strength of his geographic and astronomical studies.

Galton was profoundly influenced by the publication of Darwin's *Origin of Species* in 1859, and turned to the study of heredity. His first famous work, *Hereditary Genius* was published in 1869. In the 1870s Galton conducted experiments with sweetpeas to determine the laws that governed the inheritance of size. The question he asked himself was

> How is it possible for a whole population to remain alike in its features … during many generations, if the average produce of each couple resemble their parents? [4]

He found the answer in regression (or reversion) toward the mean. Reversion checks what would otherwise cause the dispersion of the race from its mean to indefinitely increase over the generations. Galton also discovered the relation between the standard error of estimate, the coefficient of determination, and the standard deviation of the Y variable. In his studies this meant

standard deviation of family

$$= \sqrt{1 - r^2} \times \text{standard deviation of general population.}$$

Galton published *Natural Inheritance* in 1889, which stimulated statistical work in genetics and biometrics by F. Y. Edgeworth (1845–1926), W. F. E. Weldon (1860–1906) and Karl Pearson (1857–1936). In the introduction to *Natural Inheritance*, he made the following observations about regression.

I have a great subject to write upon. It is full of interest of its own. It familiarizes us with the measurement of variability, and with curious laws of chance that apply to a vast diversity of social subjects. [5]

References

1. Helen M. Walker, *Studies in the History of Statistical Method*, Williams and Wilkins, Baltimore, MD, 1929, p. 98.
2. Jesse W. Markham and Paul V. Teplitz, *Baseball Economics and and Public Policy*, Lexington Books, Lexington, MA, 1981, p. 81.
3. Churchill Eisenhart, "The Meaning of 'Least' in Least Squares," *Journal of the Washington Academy of Sciences*, **54** (1964), 24–33.
4. Walker, op. cit., p. 103.
5. Quoted in George A. Miller (ed.), *Mathematics and Psychology*, Wiley, New York, 1964, p. 148.

CHAPTER 11

Karl Pearson

Karl Pearson (1857–1936) has been called "the founder of the science of statistics." [1] Helen Walker identifies the start of the first great wave of modern statistics with the publication of Galton's *Natural Inheritance* in 1889 and with Pearson's series of lectures on *The Scope and Concepts of Modern Science* in 1891 and his series in 1893 on *Mathematical Contributions to the Theory of Evolution*. [2] Prior to this time statistical theory had been primarily the work of astronomers and mathematicians concerned with errors of measurement.

Karl Pearson—to 1901

Pearson was the son of a Yorkshire barrister. He obtained a scholarship to King's College, Cambridge at the age of 18 and took his degree in mathematics 4 years later (being the third wrangler). For the next 5 years he traveled and studied history and social thought in Germany and lectured on related topics in England. He was a socialist and a self-described "free-thinker." In 1884 he secured an appointment in Applied Mathematics and Mechanics at University College in London. By 1890 he was approaching a critical juncture in his career, and J. B. S. Haldane imagined how Pearson would have been evaluated at this time.

> He is a first-rate teacher of applied mathematics. He is somewhat of a radical, but he is only 33 years old. He will settle down as a repectable and useful member of society, and may expect a knighthood if he survives to 60. He will never produce work of great originality, but the college need not be sorry to have appointed him. [3]

Galton's *Natural Inheritance* was published in 1889, and Pearson was inspired to broaden his horizons. He wrote later

> I interpreted ... Galton ... to mean that there was a category broader than causation, namely correlation, ... and that this new conception of correlation brought psychology, anthropology, medicine, and sociology in large parts into the field of mathematical treatment. It was Galton who first freed me from the prejudice that sound mathematics could only be applied to natural phenomenon under the category of causation. Here for the first time was a possibility, I will not say a certainty, of reaching knowledge—as valid as physical knowledge was then thought to be—in the field of living forms and above all in the field of human conduct. [4]

In 1890 Pearson secured an appointment at University College in London with freedom to chose his intellectual path. This led to the lectures on "The Scope and Concept of Modern Science," which in turn were expanded into Pearson's *Grammar of Science*, published in 1892.

The Grammar of Science, writes E. S. Pearson, was "an event in the intellectual development of many of the older generation of today." In it, Karl Pearson contrasted scientific method as a route to knowledge with the establishment of truth by authority, which he called metaphysics. He saw the objective of science as the establishment of natural laws. These are inevitable sequences of natural events that are found through the marshalling of facts and the examination of their mutual relationships. Modern science can lay claim to the support of society because its method is essential to good citizenship, its results bear on the practical treatment of social problems, and because man's quest for knowledge demands that experience be condensed into forms in which harmony exists between the representation and what is represented. Knowledge and insight are deepened when we find an arrangement of facts that "falls into sequences which can be briefly resumed in scientific formulae." [5]

Jerzy Neyman (1894–1981) wrote of the influence of *The Grammar of Science* on his generation. Neyman, a major figure in statistics in the 1930s, was then studying mathematics at the University of Kharkov in the Ukraine. He recalled:

> This was in the summer vacation of 1916; in the next session the Revolution was upon us; in our many meetings, with endless discussion on politics, *The Grammar of Science* was frequently, mentioned. Whether we joined the 'reds' or the 'whites' or stood by as skeptical onlookers, its teaching remained to influence our outlook. [6]

From 1893 to 1901 Pearson's contributions to the emerging methodology and practice of statistics are found in contributions to the *Proceedings* and the *Philosophical Transactions of the Royal Society*. Pearson's research involved the laws of heredity, but to carry out his investigations required the development and extension of statistical methods. These included the fitting of mathematical curves to the frequency distributions of observed data, the development of basic formulas in simple and multiple correlation, and the introduction of a test of goodness-of-fit of a mathematical curve, or model, to observed data. E. S. Pearson terms the goodness-of-fit test, known as chi-square, "one of Pearson's greatest single contributions to statistical theory." [7]

We now turn to this test.

Chi-Square Goodness-of-Fit Test

In the chapter on "Probability" we calculated the probabilities for lengths of a Baseball World Series between evenly matched teams. If a pair of teams is evenly matched, the probability that either wins any game is 0.5. From this basic probability, the probabilities for different lengths of the series can be calculated using accepted probability laws if it is assumed that the outcomes of the games are independent. We showed these calculations along with the lengths of series for the first 67 World Series. The data are given in Table 11-1.

The probability of a series lasting four games among evenly matched teams is 0.1250. The actual number of four-game series in repeated runs of 67 series among evenly matched teams is a binomially distributed variable with an expected value of 67(0.1250) = 8.38. In the 67 series observed there were 12 four-game series. This is more than expected for evenly matched teams, but is it enough more to cast doubt on the hypothesis of evenly matched teams? We would want to know the probability of a result departing this much (or more) from what is expected of evenly matched teams. If that probability is quite small, then our result is unusual for evenly matched teams and we might rather conclude that the teams were not evenly matched. If you throw six pennies and get six heads, this is a result that occurs with probability 0.015625 in repeated tosses of six evenly balanced coins (ones for which the probability of a head is 0.5). Results as unusual as this (six heads or six tails) have a probability of occurrence of 2(0.015625) = 0.03125. To believe in the face of the evidence of six heads that the coins are fair, you have to think that an event that occurs roughly 3 times in 100 under those circumstances just happened. That is

Table 11-1. Calculation of Chi-Square for Test of Hypothesis of Evenly Matched World Series.

Number of Games X	Number of Series f(A)	P(E)	f(E)	Chi-sq.
4	12	0.1250	8.38	1.57
5	16	0.2500	16.75	0.03
6	15	0.3125	20.94	1.68
7	24	0.3125	20.94	0.45
Total	67	1.0000	67.00	3.73

taking long odds. It is more prudent to conclude that the coins are not fair.

We have only talked about the number of four-game series. What about the rest of the results? What Pearson found was the probability distribution for an aggregate measure of discrepancy between an actual distribution and the frequencies expected on the basis of a theory, or mathematical model. The distribution is called chi-square, and is related to the sum of squared standard scores for independent draws from a normal distribution.

In the kind of applications we are showing here, chi-square is calculated from

$$\text{chi-square} = \chi^2 = \sum \frac{[f(A) - f(E)]^2}{f(E)},$$

where $f(A)$ stands for the actual frequencies and $f(E)$ stands for the frequencies expected under the theory being tested. The chi-square distribution has been tabulated for what are called different *degrees of freedom*. Once the value of chi-square has been calculated, we can consult published tables of chi-square to find the probability of a chi-square value as large or larger than the one we have. If this probability is small, doubt is cast on the validity of the theory. One convention is to compare the observed value of chi-square with the value that has only a 0.05 probability of being exceeded. Table 11-2 gives the values of chi-square that have a 0.05 probability of being exceeded.

The chi-square measure of discrepancy for the World Series data is 3.73. This value must be compared with the critical value of the chi-square distribution for the appropriate degrees of freedom. Degrees of freedom represent the number of *independent* opportunities for difference between the actual frequencies and the frequencies expected under a theory. To find the number of degrees of freedom we start with the number of classes for which we are comparing actual with expected frequencies. Then we subtract one degree of

Table 11-2. Upper 5 Percent Cut-Off Values of Chi-Square.

Degrees of freedom	Chi-sq.	Degrees of freedom	Chi-sq.
1	3.84	11	19.7
2	5.99	12	21.0
3	7.81	13	22.4
4	9.49	14	23.7
5	11.1	15	25.0
6	12.6	16	26.3
7	14.1	17	27.6
8	15.5	18	28.9
9	16.9	19	30.1
10	18.3	20	31.4

freedom for each independent way in which the two sets of frequencies are made to agree. The only sense in which agreement was forced was between the total expected and the total actual frequencies. For this reason there are $4 - 1 = 3$ independent components of chi-square, or degrees of freedom.

The chi-square measure of discrepancy for the World Series data is 3.73. The value of chi-square that has a 0.05 probability of being exceeded for three degrees of freedom is 7.81. Our chi-square is not unusual for evenly matched teams. In fact it is not far from the average chi-square for these circumstances, which is 3.0 (the degrees of freedom). The evidence does not cast doubt on the hypothesis, or theory of evenly matched teams appearing in the World Series.

In Table 11-3 we calculate the chi-square statistic for the examples that were given in Chapter 7 to illustrate the Poisson distribution. Here, two new elements occur. First is the grouping of classes when an expected frequency is less than 5.0. The second is that the degrees of freedom in these examples is the number of components of chi-square less 2 rather than the number less 1. The reason is that agreement between actual and theoretical frequencies is forced in an additional way. The Poisson mean, μ, is estimated from the data. The expected distribution is thus made to agree with the actual distribution in this respect as well as in respect to the total frequency. Two degrees of freedom are lost.

Again we consult a table of chi-square and find there is no reason to doubt the validity of the Poisson distribution as an underlying theory, or model for these three collections of data.

A final example is Quetelet's data for chest girths of Scottish soldiers that we showed in the first chapter. The expected frequencies are for a normal distribution with the same mean and standard deviation as the actual data. The chi-square statistic is 49.16, which is much larger than the chi-square value that is exceeded with probability 0.05. Something is amiss, and looking at the chi-square components provides a hint. Most of the lack of fit appears in

Table 11-3. Fit of Poisson to Several Events.

	Deaths from horsekicks			Outbreaks of war			Flying bomb hits		
X	$F(A)$	$F(E)$	Chi-sq.	$F(A)$	$F(E)$	Chi-sq.	$F(A)$	$F(E)$	Chi-sq.
0	144	139	0.18	223	216.2	0.21	229	226.8	0.02
1	91	97.3	0.41	142	149.6	0.39	211	211.4	0.00
2	32	34.1	0.13	48	51.8	0.28	93	98.6	0.32
3	11 ⎫	8 ⎫		15 ⎫	12 ⎫		35	30.6	0.63
4	2 ⎭	1.4 ⎬ 1.20		4 ⎭	2.1 ⎬ 1.47		7 ⎫	7.1 ⎫	
5		0.2 ⎭			0.3 ⎭		⎬	1.3 ⎬ 0.04	
6								0.2 ⎭	
7							1 ⎭		
Sum 280		280	1.92	432	432	2.35	576	576	1.01
	chi-sq. (0.95, 2) = 5.99			chi-sq. (0.95, 2) = 5.99			chi-sq. (0.95, 3) = 7.81		

Table 11-4. Calculation of Chi-Square for
Normal Fit to Chest Measures of Scottish
Soldiers.

Girth	$f(A)$	$P(E)$	$f(E)$	Chi-sq.
33	3	0.0007	4.02	
34	18	0.0029	16.64	0.01
35	81	0.0110	63.12	5.07
36	185	0.0323	185.34	0.00
37	420	0.0732	420.02	0.00
38	749	0.1333	764.88	0.33
39	1073	0.1838	1054.64	0.32
40	1079	0.1987	1140.14	3.28
41	934	0.1675	961.12	0.76
42	658	0.1096	628.88	1.35
43	370	0.0560	321.33	7.37
44	92	0.0221	126.81	9.56
45	50	0.0069	39.59	2.74
46	21 ⎤	0.0016	9.18 ⎤	
47	4 ⎬	0.0003	1.72 ⎬	18.38
48	1 ⎦	0.0001	0.57 ⎦	
Total	5738	1.0000	5738.00	49.16

chi-sq. (0.95, 10) = 18.31

the upper half of the distribution. The largest discrepancy occurs at 46 inches
in girth, where the actual number is nearly twice what is expected for a sample
from a normally distributed population. There are also more than expected at
45 inches but less than expected at 44 inches. Could it be that some of those
with substantial girth around 44 inches actually puffed themselves up to 45 or
46 inches? Karl Pearson observed in his lectures that some writers had
illustrated the normal law of errors with data which in fact did not fit the
normal law. In his work, *Hereditary Genius*, Galton had called attention to
Quetelet's data, citing the "marvelous accordance between fact and theory."
[8] In fairness we should add that in a sample as large as 5738 measures, even a
minor disagreement between fact and theory will be pointed out by the chi-
square criterion. But it is a disagreement nevertheless.

We hope you agree with E. S. Pearson that "chi-square was indeed a
powerful new weapon in the hands of one who sought to battle with the myths
of a dogmatic world." [9]

Karl Pearson—from 1901

In 1900 Pearson had completed an extensive paper on variation and likeness
of fraternal characteristics compared with the variation and likeness of organs

produced by the same plant in the vegetable kingdom. The paper was submitted to the Royal Society, whose referees returned a critical report. While the paper was eventually published, Pearson felt that it had been handled unfairly and that the establishment in biology did not understand and were not sympathetic to the statistical approach. At the suggestion of W. F. R. Weldon, a zoologist and naturalist who shared Pearson's convictions, it was determined to establish a new journal, which became *Biometrika*, for the publication of mathematical and statistical contributions to the life sciences. Pearson became the editor of the journal, a position he continued to hold until his death 37 years later.

In 1911 the Biometrics Laboratory and the Eugenics Laboratory of the University of London were combined into the Department of Applied Statistics with Pearson holding the title of Galton Professor. Many years earlier Florence Nightingale (1820–1910), a pioneer in applying statistics to health problems, had suggested to Galton that such a professorship be established for the statistical investigation of social problems. She had enumerated issues in education, criminology, and social services in which the effects of various existing and proposed measures were unknown, and had asked Galton to add to the list and assist her in seeing that the professorship was established. [10]

About this time Pearson began work on a biography of Francis Galton. The entire work, entitled, *The Life, Letters, and Labours of Francis Galton*, spans four volumes and 20 years of Pearson's life. At the finish of this work in 1930, Pearson wrote

> It may be said that a shorter and less elaborate work would have supplied all that was needful. I do not think so ... I have written my account because I loved my friend and had sufficient knowledge to understand his aims and the meaning of his life for the science of the future ... I will paint a portrait of a size and colouring to please myself. [11]

The result is one of the world's great biographies.

In addition to his teaching, Pearson continued to publish contributions to statistical theory, genetics and eugenics, as well as to edit *Biometrika* and to involve himself in studies of tuberculosis, alcoholism, mental deficiency, and insanity. He did not shy away from controversy. On the contrary, he regarded himself as a crusader, having written in 1914.

> ... as I grow older I feel the need for ... a species of watch-dogs of science, whose duty it shall be not only to insist on honesty and logic in scientific procedure, but who shall warn the public against appearances of knowledge where we are as yet in a state of ignorance.
>
> In many ways the trained scientific mind can warn the public ... (against quackery and dogma) ... and this is, above all, the case when the final problem turns on the interpretation of figures. To figures, in my experience, ultimate appeal is invariably made, and too often this appeal is in the inverse ratio to the power of handling them. [12]

Helen Walker concludes of Karl Pearson

> What he did in moving the scientific world from a state of sheer disinterest in statistical studies over to a situation in which a large number of well-trained persons were eagerly at work developing new theory, gathering and analyzing statistical data from every content field, computing new tables, and reexamining the foundations of statistical philosophy, is an achievement of fantastic proportions.
>
> Few men in all the history of science have stimulated so many other people to cultivate and enlarge the fields they had planted. [13]

References

1. S. S. Wilks, "Karl Pearson: Founder of the Science of Statistics," *The Scientific Monthly*, **53** (1941), 249–253.
2. Helen L. Walker, "The Contributions of Karl Pearson," *Journal of the American Statistical Association*, **53**, No. 281 (March 1958), 11–22.
3. Quoted in H. L. Walker, op. cit., p. 15 (J. B. S. Haldane was an eminent geneticist and statistician.)
4. Quoted in Egon S. Pearson, *Karl Pearson: An Appreciation of Some Aspects of His Life and Work*, Cambridge University Press, Cambridge, UK, 1938, p. 19.
5. E. S. Pearson, op. cit., p. 24.
6. E. S. Pearson, op. cit., pp. 21–22.
7. E. S. Pearson, op. cit., p. 29.
8. Francis Galton, *Heriditary Genius* (1869), reprinted by Meridian Books, New York, 1962, p. 70.
9. E. S. Pearson, op. cit., p. 31.
10. Helen L. Walker, *Studies in the History of Statistical Method*, William and Wilkins, Baltimore, MD, 1929, pp. 172–174.
11. E. S. Pearson, op. cit., p. 81.
12. E. S. Pearson, op. cit., pp. 65–66.
13. Helen L. Walker, "The Contributions of Karl Pearson," op. cit., p. 222.

CHAPTER 12

Pearson to Gosset to Fisher

A double play in baseball is a beautiful thing to behold, and an infield that can turn the double play is much to be valued. The most common double play is short-stop to second baseman to first baseman. In the early 1900s in baseball the most famous combination was Tinker to Evers to Chance for the Chicago Cubs. Our next episode in statistical methods has the heroic quality of the double play. Our team has one putout on the opposition up to now (1908). A spectacular double play is about to happen which gets us out of the inning and ready to go on the offensive. We call the play Pearson to Gosset to Fisher.

William S. Gosset (Student)

William S. Gosset (1876–1937) studied mathematics and chemistry at New College, Oxford, and in 1899 went to work for the Guinness Company as a brewer. In brewing, outcomes are susceptible to variability in materials and temperature changes, and a number of recent university graduates with scientific degrees had been taken on by the company. The methods of statistics up to this time had emphasized the normal distribution of repeated sample means (see our Chapter 8). For these applications the standard deviation of the population must be known. When statisticians of the day did not know the standard deviations of the populations they dealt with, they relied on fairly large samples. They knew that this was an approximate solution, but guessed (correctly) that for large samples (50 or more) the inexactness was not important. Many of the collections of data from that era strike us now as very large. For example, G. Udney Yule (1871–1951) used an illustrative text example of heights of 8585 males in the British Isles from an 1883 report. His discussion of correlation featured data published by Karl Pearson in 1903 on the heights of 1078 father–son pairs. [1]

The sample sizes available for control and experimentation in brewing would be small, and Gosset knew that a correct way of dealing with a small sample was needed. He first consulted Karl Pearson about this problem in 1905. Pearson informed Gosset of the current state of knowledge, which was not satisfactory. The following year Gosset undertook a course of study under Pearson. In approaching his problem Gosset employed both mathematics (in which he claimed he was inadequate) and what today we call simulation. His simulation was to draw successive samples of four measures each from a population of heights and left-middle finger measurements of 3000 criminals from an anthropometric study of the day. From each of the resulting 750 samples of four measures (say height), he then had a sample mean and sample standard deviation. Then he examined the ratios of the differences between the sample mean and the population mean divided by $S/\sqrt{n}$ rather than $\sigma/\sqrt{n}$. He called these ratios z.

$$z = \frac{\bar{X} - \mu}{S/\sqrt{n}}.$$

Working mathematically Gosset derived the distribution form for these z-ratios. He found from chi-square tests that the actual ratios from the simulation did not fit the normal but did fit the new distribution he had derived mathematically. The results were published in 1908 under the pseudonym "Student." [2]

Harold Hotelling (1895–1973) contrasts the impact of Gosset's contributions from 1908–1912 with those of the mathematician–probabilist Henri Poincaré (1854–1912), whose final edition of *Calcul des Probabilities* appeared in 1912.

In contrast to Poincaré's brilliance the contributions to probability published in 1908 and 1912 by the chemist W. S. Gosset . . . seem bumbling affairs indeed. . . . Altogether the papers of this anonymous "Student" must have seemed a pretty dismal flop to any disciple of Poincaré who might somehow have been induced to look at them. Yet "Student's distribution" is a basic tool of a multitude of statisticians who will never have any use for the beautiful but relatively inconsequential work of Poincaré in probability; and what is more important, "Student" inspired Fisher. [3]

Ronald A. Fisher

Ronald A. Fisher (1890–1962) received his B.A. degree in astronomy from Cambridge in 1912, and in this same year wrote a letter to Gosset in response to Gosset's *Biometrika* paper. The letter included a rigorous proof of Gosset's z-distribution. The proof was not published until 1923, and in 1925 Fisher published a derivation of the form in which the distribution is now employed. [4] That form is

$$t = \frac{\bar{X} - \mu}{s/\sqrt{n}},$$

where

$$s = \sqrt{\frac{\sum(X - \bar{X})^2}{n - 1}}.$$

The difference is that in S the sum of squared deviations from the sample mean is divided by the sample size, n, while in s the divisor is one less than the sample size. The value of the sample size less one is called the *degrees of freedom*. Here, the degrees of freedom are the number of independent elements in estimating the average squared deviation of the population values about the population mean. One degree of freedom is lost (from the sample size) in using the sample mean in place of the unknown population mean as a base from which to sum squared deviations.

Walker dates the second period in the development of modern statistics from the time of another paper by Fisher in 1915. We resume Fisher's story later, and in the meantime we need to turn to the use of the Student t-distribution.

Using the t-Distribution

Fisher's two great works were *Statistical Methods for Research Workers*, first published in 1925, and *The Design of Experiments*, which appeared in 1935. In *The Design of Experiments* he uses data collected by Darwin on the difference between the sizes of cross- and self-fertilized plants of the same species to bring out some points about experimental design and to illustrate the t-distribution.

Darwin had sought to determine the effect of cross-fertilization on the size of plants. Pairs of plants, one cross- and one self-fertilized at the same time and whose parents were grown from the same seed, were planted and grown in the same pot. The numbers of pairs of plants were not large because the time and care needed to carry out the experiments were substantial. Darwin's experiment had taken 11 years.

Darwin had recognized the issue of sampling variation, having written

> I may premise that if we took by chance a dozen or score of men belonging to two nations and measured them, it would I presume be very rash to form any judgment from such small numbers on their (the nation's) average heights. But the case is somewhat different with my ... plants, as they were exactly of the same age, were subjected from first to last to the same conditions, and were descended from the same parents. [5]

Darwin had sent the data for several species to Francis Galton. Galton appeared to be of two minds, commenting on the one hand about the comparison between the two varieties as "perfectly reliable," and on the other that

Table 12-1. Data and Calculations for
Differences in Height* for 15 Pairs of
Plants.

Pair	Difference	Pair	Difference
1	49	9	14
2	−67	10	29
3	8	11	56
4	16	12	24
5	6	13	75
6	23	14	60
7	28	15	−48
8	41	Total	314

*Cross minus self-fertilized, in eighths inches.

at least 50 pairs of plants would be needed "to be in a position to deduce fair
results." In any case, the methods for handling small numbers of observations
were not known, and in their absence Galton tried to get too much out of the
data by informal (and erroneous) methods.

Fisher then gives the data for the 15 pairs of plants available for one of
Darwin's experiments, *Zea mays*. The sample mean is 20.93 and the sum of
squared deviations from the sample mean is 19945. So,

$$s = \sqrt{\frac{19945}{n-1}} = \sqrt{\frac{19945}{14}} = 37.74.$$

If the sample of 15 differences were a sample from a normal distribution of
differences with a mean of zero, the Student *t*-value would be

$$t = \frac{\bar{X} - \mu}{s/\sqrt{n}}$$

$$= \frac{20.93 - 0}{37.74/\sqrt{15}} = \frac{20.93}{9.746} = 2.148.$$

The denominator quantity, $s/\sqrt{n}$, is the (estimated) standard error of the
mean, in this case a mean difference. Student's *t*-value is like a *Z*-score. We
have found that Darwin's sample mean is 2.148 standard errors from the long-
run mean that would prevail if there were no difference between heights of
cross- and self-fertilized plants of *Zea mays*.

We now need only consult a table of cut-off values of the *t*-distribution to
find whether Darwin's difference is an unusual one to get if there were really
no mean difference in the heights of plants arising from the method of
fertilization. As with chi-square, the definition of *unusual* is somewhat open,
but we will take it as a sample difference that would be exceeded in less than 5
in 100 samples taken from a population where no difference existed. Table

Table 12-2. Extreme 5 Percent Cut-Off Values for Student's t.

Degrees of freedom	t	Degrees of freedom	t	Degrees of freedom	t
1	12.706	11	2.201	21	2.080
2	4.303	12	2.179	22	2.074
3	3.182	13	2.160	23	2.069
4	2.776	14	2.145	24	2.064
5	2.571	15	2.132	25	2.060
6	2.447	16	2.120	26	2.056
7	2.365	17	2.110	27	2.052
8	2.306	18	2.101	28	2.048
9	2.262	19	2.093	29	2.045
10	2.228	20	2.086	30	2.042

12-2 gives the absolute values of t that are exceeded with probability 0.05 for different degrees of freedom. We find that the t-value that has a 5 percent chance of being exceeded (both tails) is 2.145. Darwin's mean difference for *Zea mays* just barely exceeds this. Statisticians would say that the difference is *significant* at the 0.05 level.

An alternative to the analysis we just made would be to state the 95 percent confidence interval for the population mean based on the sample results. If the t-value whose absolute level is exceeded with probability 0.05 is multiplied by the standard error of the mean, $s/\sqrt{n}$, and then the result added to and subtracted from the sample mean, the resulting range is called the 0.95 (or 95 percent) confidence interval for the population mean. This means that the method of estimating the interval will lead to correct results in 95 percent of the cases in which it is used in the long run. The reason for this is that 95 percent of sample means over the long run lie closer to the population mean than 2.145 times $s/\sqrt{n}$.

For Darwin's data the 95 percent confidence interval for the population mean difference between pairs of plants is

$$20.93 - 2.145 \, (9.746),$$
$$\text{to} \quad 20.93 + 2.145 \, (9.746),$$
$$\text{or} \quad 20.93 \pm 20.91,$$

which leads to the estimation interval of

$$0.02 \text{ to } 40.81 \text{ (eighths inches)}.$$

The fact that this interval does not quite include zero (no difference between cross- and self-fertilization) shows in a different way the same situation that we showed when we found the sample mean difference just barely "unusual" using the 0.05 definition of unusual. We conclude that Darwin's data do reflect a real difference in mean height in the underlying population.

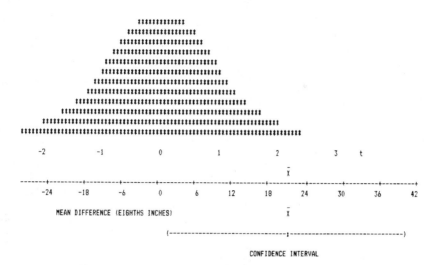

Figure 12-1. Hypothesis Test and Confidence Interval for Darwin's Experiment.

A graphic comparison of the two methods just illustrated appears in Figure 12-1. In the first method, called *hypothesis testing*, we find that the sample evidence (mean) is evidence that would occur only rarely if the hypothesis (that the population mean is zero) were true. In the second method, called *estimation*, we find a range of values that are credible for the population mean given the sample evidence. That range did not happen to include zero for Darwin's data.

If the population standard deviation were known, the 95 percent confidence interval for an unknown population mean would be established by adding and subtracting $1.96(\sigma/\sqrt{n})$ to and from the sample mean. That is, the Z-score multiple would be used instead of Student's t. A look back at Table 12-2 will show you the effect of uncertainty about the population standard deviation that is incorporated in t. The 95 percent level of t diminishes as the sample size (and degrees of freedom) increase. At 10 degrees of freedom we have 2.228 and at 30 we have 2.042. At 100 degrees of freedom we would have 1.984, and at 1000 we would have 1.962. As certainty is approached the multiple converges on 1.96, the 95 percent value for the normal Z-score.

In the next chapter we continue our discussion of regression and correlation. These are topics to which R. A. Fisher made important contributions.

References

1. G. Udney Yule, *An Introduction to the Theory of Statistics*, 5th ed., Griffin, London, 1922, pp. 83, 160.

2. Student (W. S. Gosset), "The Probable Error of a Mean," *Biometrika*, **6** (1908), 1–25.
3. Harold Hotelling, "The Impact of R. A. Fisher on Statistics," *Journal of the American Statistical Association*, **46**, No. 253 (March 1951), 37.
4. Churchill Eisenhart, "On the Transition from 'Student's' z to 'Student's' t," *The American Statistician*, **33**, No. 1 (February 1979), 8–10.
5. Quoted in R. A. Fisher, *The Design of Experiments*, 3rd edn., Oliver and Boyd, London, 1942, p. 27.

CHAPTER 13

More Regression

We ended our introduction to regression with Galton's statement that he had found a "great subject to write upon." The torch was then passed to Karl Pearson, who developed most of the mathematics of regression analysis. Then Gosset, Pearson's student, posed the problem which he ultimately solved himself by developing the t-distribution. The t-distribution allowed the error of a sample mean to be correctly stated using only the information of the sample.

Our discussion of correlation and regression was limited to describing the data at hand—that is, descriptive statistics. Based on the work of Gosset and Fisher, we are now able to go further—into the question of inference from the sample to the population. Also, this is a good time to look at the extension of regression and correlation to more than a pair of variables—one to be estimated and one to use as the basis of estimating. We will go on to the use of several variables to estimate a variable of interest.

Standard Error of the Regression Coefficient

The regression coefficient tells us what the average change in Y is for a unit increase in X. More often than not the data used to calculate this average change are a sample. When this is the case we will want to consider the standard error of the coefficient. The standard error is

$$\text{SE}(b) = \sqrt{\frac{\sum (Y - Yc)^2}{(n - 2) \sum (X - \bar{X})^2}}.$$

For the baseball runs regression, $\text{SE}(b) = 0.284$. The 95 percent confidence interval for the population regression coefficient is

$$b - t(0.975, n - 2) * \text{SE}(b) \text{ to } b + t(0.975, n - 2) * \text{SE}(b),$$
$$0.72 - 2.228(0.284) \text{ to } 0.72 + 2.228(0.284),$$
$$0.09 \text{ to } 1.35.$$

If the data are regarded as a random sample from a long-run causal system, the 95 percent confidence interval for the true regression coefficient for year t based on year $t - 1$ is 0.09 to 1.35. This is a wide range, but the sample is only $n = 12$. Because the range does not include zero, however, we can conclude with only a small risk of error, that there is real relationship between year t and year $t - 1$ team runs in the population.

Transformations

Sometimes a relationship will not be best described by an arithmetic straight line. The first place to recognize this is the scatter diagram, but it may be seen also after a straight line regression is calculated and plotted along with the data scatter. The data in Figure 13-1 are per capita income and percentage share of employment in agriculture for 20 OECD countries in 1960. From the scatter one can tell that if a straight line fits the points above $750 per capita income it will not fit the ones below that, and vice versa. The relationship, if kept strictly arithmetic, would be a curve of some kind.

It is possible of course to fit a curve to the data, and many computer programs have this alternative under titles like *polynomial regression*. An alternative, not necessarily better or worse, is to look for some transformation of one or the other variable (or both) which will make the transformed relationship linear. In Table 13-1 we show the results of two different transformations.

First, the results of the linear arithmetic regression can be seen in Table 13-1. The estimated per capita income for Turkey based on this line turns out to be negative. If Turkey is to remain in the analysis some solution must be found. We tried two transformations. The first was a square root transformation of X, and the second was a logarithmic transformation of Y. To make the data more linear we need to stretch out the lower range of X. The square root transformation works in this direction. The comparison below shows the effect.

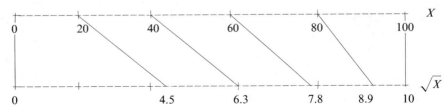

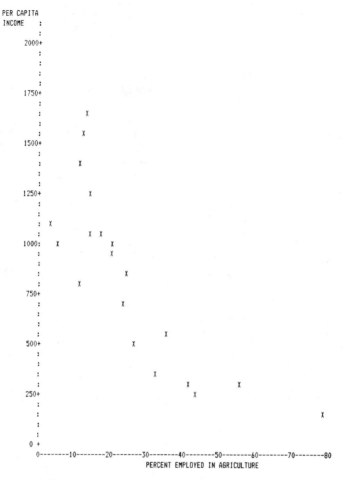

Figure 13-1. Per Capita Income and Percent Employed in Agriculture in 20 OECD Countries in 1960.

Figure 13-2 is the scatter diagram for per capita income plotted against a scale of square root of percentage employed in agriculture. The scatter is definitely more linear than before. The results show a slight improvement over the linear arithmetic regression. The square root relationship is

$$Yc = 1823.6 - 206.8\sqrt{X}$$

and this line explains 64.8 percent of the variation in per capita incomes compared with 63.2 for the arithmetic straight line. But we wind up again with a negative estimated income for Turkey.

A relationship that will not produce negative estimates and will at the same

Table 13-1. Estimated Per Capita Income from Three
Regressions.

Country	Y	X	Linear	SQRT	Expon.
Canada	1536	13	1073	1078	1048
Sweden	1644	14	1054	1050	1016
Switzerland	1361	11	1110	1138	1117
Luxembourg	1242	15	1035	1023	984
United Kingdom	1105	4	1242	1410	1396
Denmark	1049	18	978	946	894
West Germany	1035	15	1035	1023	984
France	1013	20	941	899	839
Belgium	1005	6	1205	1317	1310
Norway	977	20	941	899	839
Iceland	839	25	846	790	716
Netherlands	810	11	1110	1138	1117
Austria	681	23	884	832	763
Ireland	529	36	639	583	505
Italy	504	27	809	749	672
Japan	344	33	696	636	555
Greece	324	56	262	276	267
Spain	290	42	526	483	417
Portugal	238	44	488	452	391
Turkey	177	79	−172	−15	129
Constant			1317.9	1823.6	1584.7
Reg. Coef.			−18.8576	−206.816	0.96872
Coef. det.			0.6315	0.6477	0.6579

time increase the linearity of this data is the exponential equation

$$Yc = a(b)^X,$$

which is linear in the logarithms of Y.

$$\log Yc = (\log a) + (\log b)X.$$

The logarithmic transformation of Y will stretch out the lower values of Y, as
can be seen below.

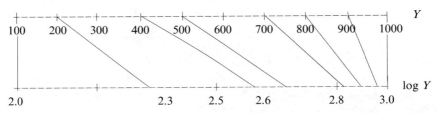

To fit the exponential regression one fits a straight line to the logarithms of Y.

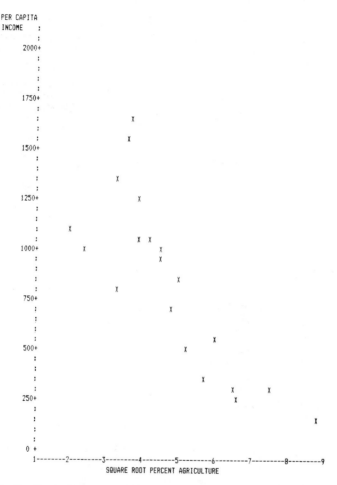

Figure 13-2. Per Capita Income and Square Root of Percent Employed in Agriculture.

Then the regression equation is used to find estimated values of log Y, and these values transformed back to natural numbers to give estimates of Y. The line is a least squares line in terms of log Y. We calculated the errors in terms of Y and then calculated the coefficient of determination as the percentage of variation in Y explained by the exponential regression. It makes no difference whether one uses base 10 or natural logarithms (base e).

The exponential regression is

$$Yc = 1584.7(0.96872)^X,$$

where X is the percentage share employed in agriculture. This regression explains 65.8 percent of the variation in per capita incomes, a little better than the square root relation. As we knew beforehand, it produces a positive per capita income for Turkey. Figure 13-3 shows the exponential regression.

The source for the country per capita income data also gives U.S. per capita income for different years adjusted to 1960 purchasing power. [1] In Table 13-2 U.S. per capita income is estimated from the exponential regression for the 20 countries (which do not include the United States), given the U.S. share of employment in agriculture for each year. Then, the standing of the actual

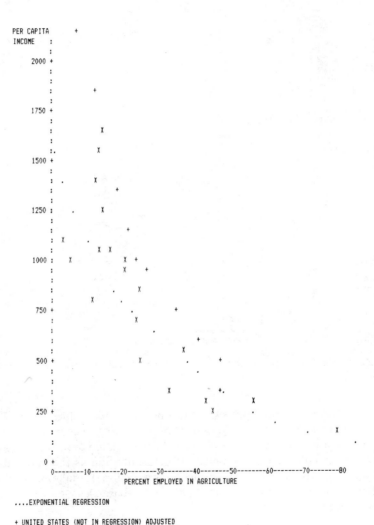

Figure 13-3. Per Capita Income and Percent Employed in Agriculture—Exponential Regression.

Table 13-2. Estimated Per Capita Income for the
United States Based on Exponential Regression.

Year	PCT AGR	Actual	Est. Y	Z-Score
1870	47	340	356	−0.06
1880	47	499	356	0.57
1890	40	592	445	0.59
1900	35	757	521	0.94
1910	28	927	651	1.10
1920	24	1050	739	1.24
1930	22	1170	788	1.53
1940	19	1364	866	1.99
1950	12	1836	1082	3.02
1960	8	2132	1229	3.61

U.S. per capita income in relation to the regression line is summed up in the Z-score for each year. The Z-score is

$$Z = \frac{Y - Yc}{\text{SE Est}}.$$

The standard error of estimate is $250, so the Z-value for 1910 is

$$Z = \frac{927 - 651}{250} = 1.10.$$

By itself, we would not judge the 1910 U.S. per capita income out of line with the relation for the other countries. One sees, however, that the Z-scores get progressively larger with time. The historic path of development in the United States does not seem to be a sample from the same causal system as is reflected by the comparative data for the other countries. And whatever the nature of the difference, it has accentuated with time.

Multiple Regression and Correlation

In the same article in which he evaluated hitters, Branch Rickey presented information on the "greatest pitchers" since 1920. [2] He agreed that ERA (earned runs average per nine innings) is an excellent overall criterion of pitching ability and included other measures that reflect parts of pitching strength. These data are given in Table 13-3. Variable (1) is just the batting average achieved against a pitcher (at bats do not include bases on balls and hit batsmen). Variable (2) measures the tendency to put men on base by bases on balls and hit batsmen. Variable (3) is a "clutch" factor—the percentage of men on base converted to earned runs, and Variable (4) is the percentage of

Table 13-3. Branch Rickey's Greatest Pitchers, 1920–1950.

Pitcher	Outcomes per 100 occasions				
	(1)	(2)	(3)	(4)	(5)
Carl Hubbell	25.1	5.3	28.0	11.2	2.98
Dizzy Dean	25.3	6.0	27.6	14.4	3.03
Lefty Grove	25.4	7.5	26.7	13.6	3.09
Grover Alexander	27.3	3.8	28.5	6.4	3.09
Dazzy Vance	25.4	7.6	28.6	16.8	3.22
Dutch Leonard	26.5	6.0	28.4	8.8	3.25
Bucky Walters	25.4	9.0	27.4	8.8	3.30
Walter Johnson	25.6	8.0	28.6	12.0	3.33
Lefty Gomez	24.3	10.5	27.3	13.6	3.34
Paul Derringer	27.2	5.2	29.8	9.6	3.46
Fred Fitzsimmons	27.2	6.7	29.8	6.4	3.54
Ted Lyons	27.5	6.6	30.1	6.4	3.67

(1) H/AB,	H = hits, AB = at bats.
(2) (BB + HB)/(AB + BB + HB),	BB = bases on balls.
(3) ER/(H + BB + HB),	HB = hit batsman.
(4) SO/(AB + BB + HB),	SO = strike-outs.
(5) ERA,	ER = earned runs.
	ERA = earned runs per nine innings.

batters who struck out. If we accept ERA as a good overall pitching measure, we might ask how the other elements of pitching contribute to it. This measurement problem can be attacked with *multiple regression.*

We will not give formulas for multiple regression for two reasons. They are cumbersome in ordinary algebra, and widely available computer programs do the calculations anyway. We will show the kind of results that computer programs give and go on from there.

In Table 13-4 we show output from a typical computer regression program. The least squares regression equation for estimating ERA on the basis of all four elements of performance is

$$Yc = -3.49 + 0.1089(1) + 0.1009(2) + 0.1158(3) - 0.0047(4).$$

The coefficients say that a one unit increase in the percentage of hits raises ERA by an average of 0.1089, a unit increase in the percentage of batsmen walked and hit increases ERA by 0.1009, and so on. Since the units and standard deviations are different, it is useful to convert to standard slopes. These are commonly known as the *beta*-coefficients. Beta coefficients are analogous to the measure r in simple regression. The standard deviation increase in estimated ERA accompanying a one standard deviation increase in the hit percentage is 0.55; the increase accompanying a one standard deviation increase in the percentage of batsmen walked and hit is 0.87, and the increase accompanying a one standard deviation change in the percentage of base

Table 13-4. Computer Regression Output.

VARIABLE	REG. COEF.	STD. ERROR COEF.	COMPUTED T	BETA COEF.
1	0.10892	0.03851	2.82825	0.55144
2	0.10090	0.01087	9.28139	0.86956
3	0.11581	0.02391	4.84279	0.59344
4	−0.00466	0.00697	−0.66871	−0.07714
INTERCEPT		−3.48912		
MULTIPLE CORRELATION		0.98432		
STD. ERROR OF ESTIMATE		0.04655		

VARIABLE	REG. COEF.	STD. ERROR COEF.	COMPUTED T	BETA COEF.
1	0.12679	0.02675	4.74025	0.64194
2	0.10287	0.01010	10.18934	0.88654
3	0.11131	0.02214	5.02772	0.57035
INTERCEPT		−3.88942		
MULTIPLE CORRELATION		0.98331		
STD. ERROR OF ESTIMATE		0.04491		

runners converted into earned runs is 0.59. From these standardized slopes it looks as if walks and hit batsmen is a more important element in determining ERA than the other variables. The strike-out rate seems rather unimportant.

In the middle columns we see the standard errors of the regressions coefficients and computed t-values. These values allow us to take into account the possible errors of random sampling in extending our conclusions to a population underlying the sample observations. In a case like the present one this population is a sort of hypothetical one—comprised of outstanding pitchers that the causal system of major league baseball between 1920 and 1950 *could* produce. Our sample includes all the outstanding pitchers that were produced—as determined by Mr. Rickey.

The degrees of freedom for the t-values for each coefficient are $n - m - 1$, where m is the number of predictors in the regression equation. In our case the degrees of freedom are $12 - 5 = 7$, and the 95 percent (two-tailed) level of t is 2.365. The first three coefficients exceed this multiple of their standard errors but not the fourth. The first three coefficients would be termed *statistically* significant at the 5 percent significance level, but not the fourth. Strike-outs are not statistically significant.

The coefficient of mutliple correlation is 0.9843. Squaring this, we find that 0.969, or 96.9 percent of the total variation in ERA is explained by the four elements of performance.

Because strike-outs have a very small beta-coefficient and are not significant, they were excluded and the analysis then run for the three remaining variables. The various measures change some, but the overall conclusions do not. For the outstanding pitchers, avoidance of walks and hit batsmen seems to be a particularly important factor in pitching performance.

The contributions of the three performance factors to the total explanatory power of the three variable equation do not add up in a simple way. The reason is that there is a pattern of correlations among the three variables. One way to "add-up" explanatory power is to start with the most powerful variable in a one-variable regression. In the present case this is the "clutch" factor, or percentage of base-runners scoring earned runs. Clutch explains 52.9 percent of the total variation in ERA, a larger percentage than either of the other two variables alone. At this point we have $100 - 52.9 = 47.1$ percent of the variation in ERA unexplained. We now ask which of the two remaining variables in combination with clutch will yield the largest percentage for explained variation. It turns out to be Variable (2), which could be called "control." When this variable is entered into a two-variable regression along with clutch, the percentage of total variation explained is 87.4. Finally, if Variable (1), the hit percentage, is included in a three-variable regression, the percentage of total variation explained rises to 96.7, the square of the multiple correlation coefficient for the three-variable equation given in Table 13-4.

We found the sequence just described by running all the possible simple and two-variable regressions in addition to the three-variable regression for explaining variation in ERA. If a larger set of possible explanatory variables were present, it would not be so easy to run all possible regressions. Computer programs called *stepwise* regression will do this kind of search for the best variable to add at each stage, add the variable, output relevant statistics, and proceed to the next stage. In fact, there are more criteria that can be employed for building the "best" equation at each stage. The one we have illustrated— additional percentage of variation explained—is generally quite serviceable, however.

History and Applications

The methods of multiple correlation and regression were worked out by Karl Pearson prior to 1910, and popularized in the next two decades of the century through texts by G. Udney Yule (1871–1951) [3] and A. L. Bowley (1869–1957) [4]. While W. S. Gosset had applied multiple regression to improving brewing quality in 1907, it is difficult to discover many applications in economics and business before World War I. A notable exception is in the estimation of demand–price relationships for corn, hay, oats, and potatoes by Henry L. Moore (1869–1958). [5] Moore is regarded as the father of econometrics.

Henry A. Wallace (1888–1965) used regression in his study of agricultural prices in 1920, and he was an enthusiastic advocate of regression methods for studies in agricultural economics. Other pioneers in the 1920s were Henry C.

Taylor, John D. Black, E. J. Working, Bradford B. Smith, Andrew Court, Louis Bean, Frederick Waugh, and Mordecai Ezekiel. Waugh's study of asparagus prices on the Boston market found that each additional inch of green raised price on the average by $38\frac{1}{2}$ cents per dozen bunches, 4 cents less per dozen bunches was received for each additional stalk (smaller stalks) in a bunch, and that the market assessed a penalty for variation in the number of stalks per bunch. Ezekiel's text, *Methods of Correlation Analysis*, was published in 1930. [6]

A book by Henry Schultz (1893–1938), *The Theory and Measurement of Demand*, published in 1938, was also important. Estimates of the demand–price relation for a number of commodities were made, and interrelations between demands found. An example is the interrelation of the demands for beef and pork. Based on annual data for 1922–1933, Schultz found that a 1 percent increase in the price of beef was accompanied by a 0.5 percent decrease in the demand for beef, and that an increase of 1 percent in the price of pork was accompanied by a 0.5 percent increase in the demand for beef. An increase of 1 percent in payroll income operated to stimulate consumption of beef by 0.33 percent. [7]

Roos and von Szeliski studied automobile demand from 1920 to 1937 to find the relation of sales to price and to income. They found that a 1 percent change in "supernumerary income" was accompanied by a change of 1.5 percent in new owner sales and a change of 1.07 percent in replacement sales. [8]

The previous two examples concerned demand relations. Paul H. Douglas and C. W. Cobb began a study of production functions in 1928, and Douglas and others studied production relations in both time series and cross-sectional data over the next 15 years. For example, consider annual data for the United States from 1900 to 1922, where P is an index of manufacturing output, L is an index of the number of wage earners employed, and C is an index of fixed capital in manufacturing. Douglas and Cobb found the relation

$$\log P = 0.0086 + 0.75 \log L + 0.25 \log C,$$

which can be restated

$$P = 1.01(L^{0.75})(C^{0.25}).$$

This means that a 1 percent increase in labor quantity was accompanied by an increase of $\frac{3}{4}$ percent in output, and a 1 percent increase in capital was accompanied by a $\frac{1}{4}$ percent increase in output. [9]

For a final example we return to baseball. For the 1968 and 1969 seasons Scully related various team inputs to team output—the win-loss percentage for major league baseball teams. [10] His final equation, which explained 88 percent of the variance in win percentage (out of 1000 points) was

$$\text{PCTWIN} = 37.24 + 0.92\,\text{TSA} + 0.90\,\text{TSW} - 38.57\,\text{NL}$$
$$+ 43.78\,\text{CONT} - 75.64\,\text{OUT},$$

TSA is the team slugging average. TSW is the team strike-out-to-walk ratio in pitching, NL is a variable which equals 1 for National League teams and 0 for American League teams, CONT is a variable which equals 1 for teams within five games of the pennant winner at the end of the season (Contenders), and OUT is a variable which equals 1 for teams 20 or more games behind the leader at the end of the season (Out of Contention). We see the net effects of slugging average and strike-out-to-walk ratio on the win percentage, and we see that teams of comparable ability on these variables do 38.57 points worse in the (tougher) National League. The net effect of being in contention is to raise the win percentage by 43.78 points and the net effect of being out of contention is to lower the winning percentage by 75.64 points beyond what would be brought about by the other variables. This regression was used as a production function in a study to determine the marginal revenue value of baseball players of different abilities.

References

1. J. Singleman, *From Agriculture to Services*, Sage, Beverly Hills, CA, 1978, p. 11.
2. Branch Rickey, "Goodby to Some Old Baseball Ideas," *Life Magazine*, August 2, 1954, 79–89.
3. G. Udney Yule, *An Introduction to the Theory of Statistics*, 1st edn. (1911), 13th edn. (1944), Griffin London, 1944.
4. A. L. Bowley, *Elements of Statistics*, P. S. King & Son, London, 1901, 1902, 1907, 1920, 1926.
5. Henry L. Moore, *Economic Cycles, Their Law and Cause*, Macmillan, London, 1914.
6. Mordecai Ezekiel, *Methods of Correlation Analysis*, Wiley, New York, 1930. The examples cited are from the third edition, *Correlation and Regression Analysis*, 3rd ed., (Mordecai Ezekiel and Karl A. Fox, eds.), Wiley New York, 1959, Ch. 25.
7. Gerhard Tintner, *Econometrics*, Wiley, New York, 1952, pp. 37–40.
8. Ibid., pp. 51–53.
9. C. F. Roos and V. von Szeliski, *The Dynamics of Automobile Demand*, General Motors, New York, 1939.
10. Gerald W. Scully, "Pay and Performance in Major League Baseball," *American Economic Review*, **64**, No. 6 (December 1974), 915–930.

CHAPTER 14

R. A. Fisher

In our Pearson to Gosset to Fisher chapter we left our historical account at Gossett's 1908 paper on what later became, at R. A. Fisher's suggestion, the Student t-distribution. We illustrated the distribution in its final form, as clarified by Fisher in 1925.

Fisher had graduated from Cambridge in 1912. For 3 years he was employed as a statistician for a London bond house and for 4 years after that he taught mathematics at the secondary school level. In 1919 the director of the Rothamstead Agricultural Experiment Station, Sir John Russell, persuaded Fisher to leave teaching for a temporary stint at the station. The station had a long series of crop yields and weather observations on Broadbalk wheat going back to 1853. The director was anxious to see what inferences could be drawn by applying some "modern statistical methods" to this data. Fisher wound up spending 14 years at Rothamstead—years which led to his major contributions to the design of experiments and their analysis.

The Analysis of Variance

We take for our first illustration of this important technique data collected by O. H. Latter in 1902 on the lengths (mm) of cuckoo eggs found in the nests of other birds. The data are reported in L. H. C. Tippett's text [1] and summarized in Table 14-1. There is variation in the mean length of eggs found in the different nests. The question might occur whether this variation is no more than might reasonably be expected in samples of this size from the same population or whether the variation is so rare from identical populations that

Table 14-1. Sample Data on Cuckoo Eggs Found in Nests of Other Birds.

	Meadow pipit	Tree pipit	Hedge sparrow	Robin	Pied wagtail	Wren	All
n	45	15	14	16	15	15	120
Mean (mm)	22.30	23.09	23.12	22.58	22.90	21.13	22.46

we had better conclude that real (population) differences exist in the mean lengths.

Let us entertain the hypothesis that the lengths of eggs in the six nests are six random samples from the same normally distributed population. Under this hypothesis there are two independent sources of variation for estimating the variance of the common population. The first is from the variation of lengths *within* the nests and the second is from the variation in lengths *among* the nests. Table 14-2 shows the results.

If the samples in the nests come from the same population, the variance of that population could by estimated from any of the samples by

$$\text{est VAR} = \frac{\sum (X - \bar{X})^2}{n - 1}.$$

The SSW row in Table 14-2 gives the within sample sums of squared deviations. Estimating the variance from the eggs in meadow pipit nests would yield

$$\text{est VAR} = \frac{37.29}{44} = 0.848.$$

But evidence about the variance is available from the other nests as well. The evidence is combined by adding the sums of squared deviations for all the

Table 14-2. Two Estimates of the Variance of Lengths of Cuckoo Bird Eggs.

Sum of squared deviations	Meadow pipit	Tree pipit	Hedge sparrow	Robin	Pied wagtail	Wren	All
Within nests (SSW)	37.29	11.38	14.85	7.03	15.96	7.74	94.25
Among nests (SSA)	1.15	5.95	6.10	0.23	2.90	26.53	42.86

Estimate from SSW: 94.25/114 = 0.827
Estimate from SSA: 42.86/5 = 8.57

types of nest and dividing by the total degrees of freedom. That is,

$$\text{est VAR}_{\text{within}} = \frac{37.29 + 11.38 + 14.85 + 7.03 + 15.96 + 7.74}{44 + 14 + 13 + 15 + 14 + 14}$$

$$= \frac{94.25}{114} = 0.827.$$

If the samples in the nests are from a common population, then the square of the difference between the mean length of eggs in the meadow pipit nests and the grand mean estimates the variance of the mean for a sample size of 45. That is,

$$\text{est VAR}(\bar{x}_{n=45}) = (22.30 - 22.46)^2 = 0.0256.$$

If the variance among means of samples of size 45 were 0.0256, then the variance among single observations (egg lengths) would be

$$\text{est VAR} = 45(0.0256) = 1.152.$$

Here again, however, evidence about the variance of single egg lengths is available, under the hypothesis being tested, from the other types of nest. In each case the estimate would be the square of the difference between the mean for the nest type and the grand mean multiplied by the sample size. The evidence from the among nest types variation is combined by adding these sums of squared deviations and dividing by the total degrees of freedom represented. The degrees of freedom are one less than the number of nest types.

$$\text{est VAR}_{\text{among}} = \frac{42.86}{5} = 8.57.$$

What Fisher worked out was the probability distribution for the ratio of two independent estimates of the variance of the same normally distributed population. Later it was called the F-distribution in honor of Fisher by George W. Snedecor (1881–1974), founder of the Statistical Laboratory at Iowa State University. It is customary to put the estimate of variance from the within samples source in the denominator of the ratio. The F-ratio for the cuckoo eggs data is

$$F = \frac{8.57}{0.827} = 10.4.$$

There is a different F-distribution for each combination of numerator and denominator degrees of freedom. For 5 degrees of freedom in the numerator estimate and 114 in the denominator estimate, the upper 5 percent value of F is 2.30 and the upper 1 percent value is 3.20. This tells us that if the two sources of variation were estimating the same normal population variance, then only 5 out of 100 repeated pairs of estimates would be expected to produce F-ratios exceeding 2.30. Only 1 in 100 would be expected to exceed an F-ratio of 3.20. The inference here is that the reason our F-ratio exceeds these cut-off values is that the two sources of variation *were not estimating the same variance*. The high F-ratio (large estimate from among samples) is attributed to the likelihood that we were sampling from populations with different (rather than equal) means. Thus, we infer that there was a source of variation affecting the among sample sums of squares other than errors of random sampling. The added source of variation would be present if the eggs in the different nest types came from populations with different means. We conclude that the mean length of eggs laid by cuckoo birds in nests of other birds differs according to species.

This cuckoo-egg example has been criticized by Lancelot Hogben (1895–1975) on the grounds that, while the study was published in *Biometrika* in 1902, it provided no new knowledge to the naturalist. That cuckoo eggs are peculiar to the locality where found was already known in 1892. A later study by E. P. Chance in 1940 called *The Truth About the Cuckoo* demonstrated that cuckoos return year after year to the same territory and lay their eggs in the nest of a particular species. Further, cuckoos appear to mate only within their territory. Therefore, geographical subspecies have developed, each with a dominant foster-parent species, and natural selection has ensured the survival of cuckoos most fitted to lay eggs that would be adopted by the foster-parent species. [2]

R. A. Fisher—from 1925

The analysis of variance was introduced by Fisher in 1924. In 1925 the first edition of Fisher's *Statistical Methods for Research Workers* appeared. Over the next 25 years the book changed the teaching and practice of statistics. Harold Hotelling (1895–1973) identifies Fisher's contributions as threefold. First were his contributions to the theory of estimation. Fisher insisted on clarification in concept and symbols between sample statistics and population values. He suggested alternative criteria for deriving estimates of population values from the statistics at hand. He emphasized the derivation of exact distributions for testing hypotheses. The F-distribution is a case in point. Finally, Fisher invented and popularized the analysis of variance. Hotelling summarizes Fisher's standing by the phrase "before Fisher was Karl Pearson, and before him were Gauss and Laplace, the real cofounders of the double science of statistics and probability." [3]

In 1931 Fisher spent a summer teaching at Iowa and Minnesota. A brochure by George W. Snedecor on the analysis of variance in 1934 helped spread the gospel among agricultural researchers in the United States. In 1936 Fisher again visited Iowa State. In 1937 Snedecor's text, *Statistical Methods*, appeared. This book introduced a generation of statisticians to Fisher's methods. In the 1960s it was one of the most cited books, according to *Science Citation Index*. [4] By 1939 Fisherian experimental design and data analysis were being put to use by agronomists world-wide. Frank Yates, his colleague, coauthor, and successor at Rothamstead, wrote of Fisher's methods in 1964

> The recent spectacular advances in agricultural production in many parts of the world owe much to their consistent use. They certainly rank as one of Fisher's greatest contributions to practical statistics, and have introduced a certainty of touch into well-designed experimental work that is the envy of statisticians faced with the interpretation of nonexperimental data. [5]

In 1943 Fisher was appointed Professor of Genetics at Cambridge. He had been Professor of Eugenics at London since leaving Rothamstead in 1933. His contributions to genetics, wholly aside from statistical methods, rank him among the leading scholars of his time in that area. [6]

Experimental Design

While we used analysis of variance to answer the question about population differences in the mean lengths of eggs laid by cuckoos in nests belonging to other species, the data certainly did not represent a controlled experiment. One would suppose that Latter made the measurements of eggs as he encountered them on field trips. We acted as if the eggs observed were a random sample of the eggs laid by cuckoos on the nests of each species, but there was no positive assurance of that. Indeed, Hogben's criticism is that there was not a designed experiment with some objective in view.

Our example of a designed experiment will follow an example given by W. J. Youden, drawn from the work of Mavis Carroll, statistician in the Central Laboratories of the General Foods Corporation. In testing food products for palatability, General Foods employed a seven-point scale for tasters varying from -3 (terrible) to $+3$ (excellent) with 0 representing "average." Their standard method for assessing the palatability of an experimental variety of a product was to conduct a taste test with 50 persons—25 women and 25 men. [7], [8]

The experiment reported here involved the effects on palatability of a coarse versus fine screen and a low versus high concentration of a liquid component. Coarse and fine screens were employed with both low and high liquid concentrations, with four groups of 50 consumers each tasting the results of one of the experimental combinations. Tasters were recruited from local churches and club groups, and were assigned randomly to the four treatment combinations as they were recruited. The results, in terms of total scores for the 50 tasters in each group, were as shown in Table 14-3.

The first principle introduced by Fisher is that of random assignment of experimental material to treatments. [9] With four treatments this could be done by flipping two coins for each subject. Each possible result (head–head,

Table 14-3. Results of Food Palatability Experiment.

		Low liquid		High liquid
Course screen	Rep. 1	35	Rep. 1	24
	2	39	2	21
	3	77	3	39
	4	16	4	60
Fine screen	Rep. 1	104	Rep. 1	65
	2	129	2	94
	3	97	3	86
	4	84	4	64

head–tail, tail–head, and tail–tail) assigns the individual to a particular treatment group. When 25 persons of the same sex have been assigned to a group, assignment is to one of three remaining groups, and so on. The purpose of random assignment is to insure that any variation not associated with treatments is truly random, or experimental error and not the result of some unknown rules of assignment.

The second principle is illustrated by the use of replications (Rep.) in the experiment depicted above. The replications, or repeat exposures of additional groups of tasters to the experimental products represent a possible source of variation over time in the results. For example, it could be that batches of the experimental product made up later were not the same in regard to factors that influence palatability but were not designed into the test. Since results are available for the separate replications, it is possible to eliminate such effects from what would otherwise be an inflated "experimental" error. The principle is that all variability not attributed to treatments does not have to inflate experimental error. In the interest of simplicity, we will not separate out the error attributable to replications, but it can be readily done.

The third principle is the balanced design featuring all combinations of treatments in one analysis. This is called a *factorial* experiment. When introduced by Fisher, the factorial experiment went against the prevailing philosophy of investigating one question at a time. Yates quotes Fisher as saying

No aphorism is more frequently repeated in connection with field trials, than that we must ask Nature few questions, or ideally, one question at a time. The writer in convinced that this view is wholly mistaken. Nature, he suggests, will respond to a logical and carefully thought-out questionnaire; indeed, if we ask her a single question, she will often refuse to answer until some other topic has been discussed. [10]

The balance in the experimental design consists of equal numbers of observations in each treatment group, and, in the case of factorial design, the use of each level of one treatment factor with each level of every other treatment factor. This produces *orthogonality*, or independence among treatment effects. This means that portions of variation can be unambiguously assigned to different treatments—in our case, screen type and liquid level. The advantages of a coordinated statistical plan versus one-factor-at-a-time experiments were portrayed as in Figure 14-1 in an article on design of experiments. [11]

Table 14-4 gives mean totals per 50 tasters by treatments. By now we should know what questions to ask. We want to know if the difference in palatability between fine and coarse screen (90.375 − 38.875) and if the difference between low and high liquid level (72.625 − 56.625) exceed what might be reasonably expected to occur on the basis of chance.

A two-way analysis of variance is available. It assumes that there is a constant effect of screen type over both liquid levels and a constant effect of liquid level over both types of screens. Under this assumption, plus an assumption of normality of the distribution of palatability scores within any screen type–liquid level combination, F-ratios can be derived for tests of the difference between screen type means and between liquid level means. Rather than show you analysis of variance computations, we will show you the equivalent analysis using regression. [12]

The regression solution for the analysis of variance just described is accomplished by using [0, 1] variables to identify the treatment combinations. We used the variable $X(1)$ for screen type, with $X(1) = 0$ standing for the coarse screen and $X(1) = 1$ for the fine screen. We used $X(2) = 0$ for the low liquid level and $X(2) = 1$ for the high liquid level. Each palatability value (Y) is associated with a value of $X(1)$ and a value of $X(2)$. For example, the observations under fine screen–low liquid level all have $X(1) = 1$ and $X(2) = 0$. We will obtain a regression equation

$$Yc = a + b(1) * X(1) + b(2) * X(2).$$

The coefficient $b(1)$ will estimate the difference in palatability between fine and coarse screens and $b(2)$ will estimate the difference in palatability between high and low liquid levels. The significance of the coefficients as indicated by the Student t-statistic is the same as the significance of the F-ratio for the indicated treatments when there are only two levels of the treatments. The regression turns out to be

$$Yc = 46.875 + 51.5 * X(1) - 16.0 * X(2),$$

and the t-statistic for the first coefficient is 5.20 and the second is −1.62. The degrees of freedom for t are $n - 3$, so the 0.05 significance level is 2.120. It is safe to conclude that a real difference exists for screen type, but the liquid level difference is questionable.

A close look at Table 14-4 might cause you to question the assumption made in our analysis that the effect of screen type is the same for both liquid levels. The sample difference for the coarse screen is $36.00 - 41.75 = -5.75$, while for the fine screen the sample difference between high and low liquid level is $77.25 - 103.50 = -26.25$. When the effect of one factor, or treat-

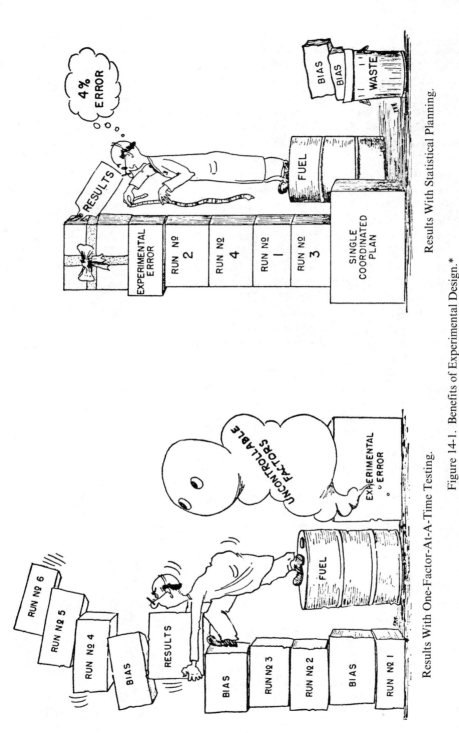

Results With Statistical Planning.

Results With One-Factor-At-A-Time Testing.

Figure 14-1. Benefits of Experimental Design.*

* Reproduced, with permission, from B. B. Day, "The Design of Experiments in Naval Engineering," *The American Statistician*, **8** (April–May, 1954), pp. 11, 13.

Table 14-4. Treatment Means for Palatability Experiment.

	Low liquid	High liquid	Total
Coarse screen	41.75	36.00	38.875
Fine screen	103.50	77.25	90.375
Total	72.625	56.625	64.625

ment, itself depends on another factor, or treatment, there is said to be *interaction* between the treatment effects. The factorial experiment with replication allows identification and test of interaction effects. In the regression approach for our example all we need do is include a product variable, $X(3) = X(1) * X(2)$. What this does is to provide the estimating equation

$$Yc = a + b(1) * X(1) + b(2) * X(2) + b(3) * X(1) * X(2).$$

When this is rewritten as

$$Yc = a + [b(1) + b(3) * X(2)] * X(1) + b(2) * X(2),$$

we see that a different change will be associated with $X(1)$ when $X(2) = 1$ than when $X(2) = 0$. With this regression we obtain

$$Yc = 41.75 + [61.75 - 20.5 * X(2)] * X(1) - 5.75 * X(2),$$

in which only the coefficient $b(1)$ is significant at the 0.05 level. We then re-ran the analysis without $b(2)$. The result was the regression

$$Yc = 38.875 + [64.625 - 26.25 * X(2)] * X(1),$$

in which the Student t-value for 64.625 is highly significant and the t-value for -26.25 is significant at 0.10 but not at 0.05. It would seem that there is no overall effect of liquid level but liquid level may affect the palatability of screen type so that a low liquid level with fine screen is most palatable.

Postscript

This has been only the briefest introduction to experimental design and Fisher's contributions to statistical methods. It brings our account up to the time of Helen Walker's "third wave" of modern statistical activity, which she dates as beginning around 1928 with the publication of certain papers by Jerzy Neyman and E. S. Pearson, the son of Karl. The controversy between Fisher and Neyman–Pearson at times was quite heated. Fisher was a fighter rather than a modest and charitable academic. In writing of Fisher in 1967, Neyman reminds his readers that he continues to maintain that some of Fisher's conceptual developments (not described in our chapter) were erroneous. But the purpose of his summary of Fisher's work was to emphasize his personal view that Fisher's record was "second to none." [13]

References

1. L. H. C. Tippett, *The Methods of Statistics*, 4th edn., Wiley, New York, 1952, p. 176.
2. Lancelot, Hogben, *Chance and Choice by Cardpack and Chessboard* Parrish, London, 1955, p. 559.
3. Harold Hotelling, "The Impact of R. A. Fisher on Statistics," *Journal of the American Statistical Association*, **46**, No. 243 (March 1951), 35–46.
4. T. A. Bancroft, "George W. Snedecor, Pioneer Statistician," *The American Statistician*, **28**, No. 3 (August 1974), 108–109.
5. F. Yates, "Sir Ronald Fisher and the Design of Experiments," *Biometrics*, **20** (June 1964), 307–315.
6. Jerzy Neyman, "R. A. Fisher: An Appreciation," *Science*, **156** (1967), 1456–1460.
7. W. J. Youden, "Factorial Experiments Help Inprove Your Food," *Industrial and Engineering Chemistry*, **49**, No. 2 (February 1957), p. 85A.
8. Elisabeth Street and Mavis G. Carroll, "Preliminary Evaluation of A Food Product," in *Statistics: A Guide to the Unknown* (Judith M. Tanur ed.), Holden-Day, San Francisco, 1972, pp. 220–228.
9. These principles are derived from M. S. Bartlett, "R. A. Fisher and the Last 50 Years of Statistical Methodology," *Journal of the American Statistical Association*, **6**, No. 310 (June 1965), 395–409.
10. Yates, op. cit., p. 312.
11. B. B. Day, "The Design of Experiments in Naval Engineering," *The American Statistician*, **8** (April–May 1954), 7–12, 23.
12. In a letter to Student in 1923 Fisher showed the equivalence of least squares regression and analysis of variance.
13. Jerzy Neyman, op. cit., p. 31.

CHAPTER 15

Sampling: Polls and Surveys

In an article called "Sampling in a Nutshell," Morris Slonim has us deal with a population of mixed nuts. [1] The mixture has one-half peanuts, one-third filberts, and one-sixth walnuts. The mean weight of the nuts in the mixture is 23 mg and the standard deviation is 17.2 mg.

Suppose there are a large number of nuts and that they have been thoroughy mixed and placed in a large container. Think of the container as a small barrel that you might find in a country store. A sample of 400 nuts is to be taken from the barrel—about the number that would fit in a good-sized dipper.

Sampling theory says that the chances are 95 in 100 that the mean weight of the nuts in the dipper will be within two standard errors of the mean weight of the nuts in the barrel. That is, the probability is 0.95 that the sampling error will not exceed

$$2.0(17.2/\sqrt{400}) = 2.0(0.86) = 1.72.$$

The standard error of the mean is 0.86. The probability is less than 0.05 that the sample mean will differ by more than $2(0.86) = 1.72$ from the population mean.

If we did not know the population mean or the population standard deviation and wanted to use the sample to estimate the mean weight of the nuts in the barrel, the 95 percent confidence interval for the population mean will be virtually

$$\bar{X} - 2.0(s/\sqrt{n}) \quad \text{to} \quad \bar{X} + 2.0(s/\sqrt{n}),$$

because the Student t-multiples for large sample sizes are nearly the same as the normal distribution multiples.

Both of the statements just made are summaries of what previous chapters have tried to make clear.

Proportions

Suppose we record a 1 when we encounter a peanut and a 0 otherwise. The occurrence of 1s in a series of random draws from the population is a Bernoulli process with a constant probability of success on each draw. Thus, the population mean is the underlying probability of a 1, $P(1)$, for the process. The standard deviation of the values (1s and 0s) in the population can be found from

$$\sigma = \sqrt{P(1 - P)}.$$

If we consider peanuts, filberts, and walnuts in turn, the mean and standard deviation of the $(1, 0)$ populations are

Mean	Standard deviation
Peanuts $P = \frac{1}{2} = 0.50$	$\sqrt{P(1 - P)} = \sqrt{0.50(1 - 0.50)} = 0.50$
Filberts $P = \frac{1}{3} = 0.33$	$\sqrt{P(1 - P)} = \sqrt{0.33(1 - 0.33)} = 0.47$
Walnuts $P = \frac{1}{6} = 0.17$	$\sqrt{P(1 - P)} = \sqrt{0.17(1 - 0.17)} = 0.37$

Let the number of 1s in a random sample of size n be r. Then, the proportion of 1s in the sample is $r/n = p$. The probability distribution of the proportion can be approximated by a normal distribution with a mean of P and a standard deviation equal to $\sqrt{P(1 - P)/n}$ if nP and $n(1 - P) > 5$. The standard score for a particular value of $p = r/n$ will be

$$Z = \frac{r/n - P}{\sqrt{P(1 - P)/n}}.$$

If we use $\text{SE}(p)$ for the standard error of the proportion, we have

$$Z = \frac{p - P}{\text{SE}(p)}.$$

Consider the probability distribution for the proportion of peanuts in a sample of $n = 400$ of the mixed nuts. The chances are a little better than 95 in 100 that the sample proportion will fall within two standard errors of the true proportion, or within

$$2 * \text{SE}(p) = 2\sqrt{\frac{P(1 - P)}{n}} = 2\sqrt{\frac{0.50(1 - 0.50)}{400}}$$

$$= 2(0.025) = 0.050.$$

The chances are only around 5 in 100 that the sample proportion of peanuts will vary by more than 0.05 from the population proportion. The probability is 0.95 that the dipper will contain between 45 and 55 percent peanuts.

Consider now the walnuts. The population proportion is $\frac{1}{6} = 0.167$. The standard error of the proportion is

$$SE(p) = \sqrt{\frac{0.167(1 - 0.167)}{400}} = 0.0186,$$

and the chances are about 95 in 100 that the sample proportion of walnuts will lie between

$$0.167 - 2(0.0186) \text{ and } 0.167 + 2(0.0186)$$
$$= 0.167 - 0.0372 \text{ to } 0.167 + 0.0372$$
$$= 0.1298 \text{ to } 0.2042.$$

If we have observed r successes in a random sample and want to establish a confidence interval for the population proportion, the interval, as long as $r \geq 50$, can be approximated by

$$p - Z * \sqrt{\frac{p(1 - p)}{n}} \quad \text{to} \quad p + Z * \sqrt{\frac{p(1 - p)}{n}}.$$

Here, Z is the standard score multiple. The product of $p(1 - p)$ based on the sample is used as an estimate of what $P(1 - P)$ would be for the population just as s is used as an estimate of σ when we deal with means.

Coefficient of Variation

Peanuts, filberts, and walnuts are not very similar in size. Suppose the means and standard deviations of weights of the nuts in the barrel are as follows:

	Mean	Standard deviation
Peanuts	9.0 mg	1.6 mg
Walnuts	28.0	2.8
Filberts	55.0	9.8

Are the weights of the peanuts more uniform than the walnuts? Their standard deviation is only 1.6 compared with 2.8 for the walnuts. Relative to their average weight, the peanuts vary more than the walnuts. They are less uniform. The standard deviation for peanuts is 17.7 percent of the mean, while for walnuts the standard deviation is only 10 percent of the mean. This comparative measure is called the *coefficient of variation*. In symbols, the coefficient of variation is

$$CV = \frac{\sigma * 100}{\mu}$$

for a population, and

$$CV = \frac{s * 100}{\bar{x}}$$

for a sample.

If weights of the peanuts and walnuts are normally distributed, we could say that 95 percent of the peanuts are within 2(17.7) = 35.4 percent of their mean, while 95 percent of the walnuts are within 2(10.0) = 20.0 percent of their mean. In this sense the walnuts weights are less dispersed.

Random Numbers

We have swept something under the rug (peanuts, no doubt) by telling you that the nuts in the barrel were *thoroughly mixed.* You might have wondered just what that meant. It meant that, except for chance variation, there was no difference in a cupful of nuts (types and weights) taken from one part of the barrel than from another. The probabilities of different nuts and the distributions of weights are constant throughout the barrel. When this is the case, we can sample the barrel anywhere and the sampling theories (probabilities, confidence intervals, etc.) shown here will apply.

Thorough mixing is an ideal that can be approached in situations like mixing nuts. Even this is not as easy as you might think. For example, in 1970 the order in which men were drafted for the military was determined by the order in which capsules containing dates of birth were drawn from a basket. It developed after the drawing that birthdates in months toward the end of the year were drawn on the average earlier than birthdates toward the beginning of the year. This would not be expected in a truly random drawing. The capsules had been put into the cage in monthly order—the early months first. The mixing had not been thorough, with the result that the last dates in tended to be the first ones drawn out. In the 1971 draft the dates were put in one container and the numbers 1 to 365 were placed in another container. Both containers were mixed and the draws made in pairs to associate a birthdate with a draft order. [2]

From the draft example we see that thorough mixing is not easy. Let us shift to sampling a human population. Suppose we are interested in the incomes of doctors, staff, and unskilled workers in the health-care system of a large city. How is the equivalent of thorough mixing achieved? The answer is with the use of random numbers. Table 15-1 is a section from a large table of random numbers. It appears to be a sequence of the digits 0 to 9 (arranged in pairs) in no systematic order or pattern and with equal frequency of each digit. That is the intention.

We would need a list of names (or other ID) of all the health-care personnel in the community. Then the list would be numbered sequentially. Suppose

Table 15-1. Segment of a Table of Random Numbers.

24	51	21	05	83	13	30	50	03	31	18	66	51	51	63	78	88	90
90	15	91	49	71	02	36	37	51	98	45	45	72	73	13	95	62	25
00	68	34	96	94	29	00	17	51	04	07	67	81	86	74	89	66	08
66	02	11	55	44	89	80	72	34	58	39	30	85	80	12	86	88	37
50	68	61	44	81	63	77	88	99	86	77	39	03	31	04	60	36	78
28	65	46	21	52	74	78	46	20	13	71	19	22	65	43	87	67	62
03	33	64	85	73	99	56	97	57	36	46	75	87	64	80	37	36	53
76	32	84	20	13	71	60	85	81	22	55	79	45	32	41	74	55	47
25	71	04	93	00	91	03	97	37	27	78	01	88	51	28	00	32	27
04	30	24	49	52	67	12	73	17	74	91	37	37	70	83	73	33	08

there are 10,000 names, which we have numbered from 0000 to 9999. A random sample would be chosen by reading successive four-digit numbers from a table of random numbers. If we started in the section reproduced in Table 15-1, the random sample would include the names numbered 2451, 2105, 8313, 3050, and so on for n names. The mixing that was done by shaking the barrel of nuts is being done in this case by the random number table. It is being done more thoroughly than would happen by trying to mix pieces of paper in an urn.

One of the first series of random numbers was published by L. H. C. Tippett in 1927. Tippett worked for many years with the British Cotton Industries Research Association as a consultant on quality control and improvement of processes. The 41,600 digits in Tippett's table were taken from the figures on areas of parishes contained in British census returns (omitting the first and last two digits in each case). Fisher and Yates in 1943 published 15,000 numbers taken from the fifteenth to nineteenth places in 20-place tables of logarithms. In 1939 Kendall and Babington–Smith generated a table of 100,000 random numbers by means of an electronic machine that simulated the tossing of a ten-sided cylinder—one side for each digit 0 through 9. Nowadays random numbers are generated by computer routines. The development and testing of better random number generators is itself a rather special area in numerical analysis.

W. J. Youden tells a story of Tippett being introduced by Edward U. Condon, Director of the Bureau of Standards. Condon illustrated the idea of the chance that random events will form a meaningful sequence by the possibility that a monkey banging on a typewriter might produce a Shakespeare play. Then he introduced Tippett by remarking that he had written a book that really could have been produced by a monkey. [3]

Other stories are told about typewriting monkeys. Warren Weaver reports an address before the British Association for the Advancement of Science (abbreviated the British Ass.) in which the speaker said that "if six monkeys were set before six typewriters it would be a long time before they produced by mere chance all the books in the British Museum; but it would not be an infinitely long time." Weaver then quotes a poem by Lucio in the Manchester Guardian about the address. The second two stanzas are

> Give me half a dozen monkeys
> Set them to the lettered keys
> And instruct these simian flunkies
> Just to hit them as they please
> Lo! The anthropoid plebians
> Toiling at their careless plan
> Would in course of countless aeons
> Duplicate the lore of man
>
> Thank you, thank you, men of science
> Thank you, thank you British Ass!
> I for long have placed reliance
> On the tidbits that you pass
> And this season's nicest chunk is
> Just to sit and think of those
> Six imperishable monkeys
> Typing in eternal rows [4]

Sampling Methods

What is called a *simple random sample* can be achieved if we are able to number all the elements in a population and then select the sample with the aid of a table of random numbers. When we are able to do this, known formulas can be applied to gauge the size of sampling error. Statistical surveys in practice are not always simple random samples. There are other *probability* sampling methods that are commonly used. Surveys often use these methods in combination.

Stratified Sampling. When the variable being studied differs among identifiable segments of the population, sampling error can be reduced by sampling the different segments, called *strata*, separately. Then the results are combined to make an estimate for the entire population.

A case in point is the mixed nuts population. Peanuts, filberts, and walnuts differ in average weight. If we were called on to estimate the mean weight for all nuts in the barrel based on a sample of 400, we could improve on the simple unrestricted sample discussed earlier. Knowing the proportions put into the mixture, we could take a sample of

$$\tfrac{1}{2} \times 400 = 200 \text{ peanuts,}$$
$$\tfrac{1}{3} \times 400 = 133 \text{ filberts,}$$
$$\tfrac{1}{6} \times 400 = 67 \text{ walnuts.}$$

Then we would find the sample mean for each subgroup and combine them to get the estimated population mean, or

$$\bar{X}(\text{all}) = 0.50\bar{X}(\text{P}) + 0.33\bar{X}(\text{F}) + 0.17\bar{X}(\text{W}),$$

where P, F, and W stand for peanuts, filberts, and walnuts. The sampling error for the stratified sample will be smaller than the error of an unrestricted random sample of the same total size because the stratified sampling error depends on *within* strata standard deviations rather than the standard deviation of all the elements mixed together.

A naturally occurring example of stratified sampling is provided by estimating the weight of a trawler's haddock catch. [5] The haddock are routinely sorted into groups by size (small, small-medium, large-medium, and large) and counted. The weight of the catch is estimated by first finding mean weights in samples of the fish in each category. Then these are combined for the overall average weight using the proportions of fish by weight classes found from the count. The estimated weight of the entire catch is the overall sample mean weight times the count of fish caught. The error of this estimate will be much less then the error from a sample of an equal number of unsorted fish.

Unlike the nuts example, the fish-weight example did not call for subgroup samples in proportion to the counts of total fish in the groups. Actually the counts of fish could be being completed while the weighing of the samples was going on. Then the proportions from the counts are applied in the overall mean calculation. Samples where the numbers in the strata are proportionate to known universe counts are common, however. They are called *proportionate stratified random samples.*

Cluster Sampling Often the elements being sampled come in clusters which are more easily identified and counted than the elements themselves. Then it is appropriate to sample the clusters, but the sampling error for a given number of final elements will be larger than for an unrestricted sample of the same number of elements.

Cochran, Mosteller, and Tukey give the example of sampling the leaves of a certain variety of tree in a forest to find the proportion infested by a parasite. [6] Certainly one cannot list and number leaves, but one can conceive of listing and numbering trees in order to use a random number table to select ones for the sample. A sample of 1000 leaves could be collected by sampling 10 leaves on each of 100 trees. The trees are the clusters. The 10 leaves on each tree would have to be chosen in a psuedo-random fashion (not all from one branch, etc.). Even then, if the property of infestation were not thoroughly mixed in each tree, the cluster sample would lose statistical effectiveness. The worst case is when each tree is either 0 or 100 percent infested. In that case the 1000 leaves are equivalent to only 100 leaves selected directly at random.

Generally, the situation with clusters will be somewhere between the extremes of thorough mixing and extreme differences. Samples of households in a city-wide survey are often accomplished by first numbering and selecting blocks and then selecting households from the chosen blocks. Blocks are not ideal clusters because households within blocks tend to be alike in income, ethnicity, etc. But their use is usually practical and economical.

Systematic Sampling. A common selection procedure is to chose every *n*th unit in a list. Examples are every hundredth customer in a credit file and every fifth household in a city block. This method is the same as random sampling only when the population units are thoroughly mixed thoughout the listing (order) in regard to the variables under study. Even when this assumption is not made, systematic sampling may be useful if proper alternatives to random sampling formulas for sampling error are used. The selection of every *n*th unit is often easier to carry out than numbering and selection using random number tables.

Election Polls

Election polls are a statistical activity visible to the public. Like the weather, how candidates are doing and who will be elected is a subject of apparently endless fascination. One of the earliest polls reported in the United States was in 1824 when the *Harrisburg Pennsylvanian* reported the results of a straw poll taken in Wilmington, Delaware. The poll showed Andrew Jackson far in the lead among presidential aspirants. [7]

A straw poll is one in which just about anyone can respond, or "vote." There is no control over who is in the sample. This kind of poll was the rule until 1933. The largest straw polls were conducted by the *Literary Digest*. In 1920 the *Literary Digest* mailed 11 million ballots, and in 1932 the mailing had increased to 20 million, of which 3 million were returned. The mailing lists were for owners of telephones and automobiles. The 1932 *Literary Digest* poll came within nine-tenths of a percentage point of Roosevelt's winning vote percentage.

In 1936 the *Literary Digest* mailed 10 million postcard ballots, and about 2 million were returned. Alfred Landon received 57 percent of these "votes" and Franklin Roosevelt 43 percent. Roosevelt actually received 62.5 percent of the major party vote in the election. The *Literary Digest* poll was nearly 20 percentage points in error!

The accepted explanation for the *Literary Digest*'s success in 1932 and failure in 1936 is that a bias in the mailing lists toward higher income levels that was unimportant in 1932 became critical in 1936 because people voted more along economic lines. In any event there was virtually no methodology worthy of the name in straw ballot polls.

In 1936, a "new generation" of pollsters, led by Elmo Roper, Archibald Crossley, and George Gallup, used what they called "scientific" polls to correctly forecast the Roosevelt victory. [8] These new pollsters used smaller numbers of interviews and put great emphasis on achieving a sample that was a "miniature replica" of the voting population. Gallup began his national polling with around 60,000 interviews, but soon was relying on closer to 10,000. Interviews were conducted in person to cut down the biases in voluntary response to straw polls. The technique of the day was "quota sampling." In an attempt to insure a "representative" sample, interviewers would be instructed to fill quotas for the sexes, income groups, political parties, and so on. Except for these quotas there were few controls.

The new polls were successful in predicting Roosevelt's victories in 1940 and 1944, though some might think this about as challenging a task as predicting that Joe Louis would win another heavyweight title fight. In 1948, Harry Truman was generally given little chance in the election against Thomas E. Dewey. The October issue of *Fortune* concluded that "Dewey will pile up a popular majority only slightly less than that accorded Mr. Roosevelt in 1936 when he swept by the boards against Alf Landon!" [9]

The 1948 election was very close. There was some evidence that Truman's vigorous campaign had cut down on Dewey's early advantage, but it was ignored at the time because "everyone knew better." Some of the polls did not conduct late surveys that might have underscored the shift. Post-election surveys suggested that one-in-seven voters made the decision about whom to vote for in the final two weeks of the campaign.

Truman won with 49.5 percent of the presidential vote. Dewey had 45.1, Thurmond 2.4, and Wallace 2.4, and other candidates 0.6 percent. Crossley had estimated Truman's vote at 44.8, and Roper had estimated it at 37.1 (Roper had discontinued polling early). Gallup had estimated Truman's vote at 44.5 percent of the major party vote, while he actually got 49.8 percent. Only Roper's error appeared out of line with errors made by the same pollsters in 1936 when they made their reputations. The polls had consistently overestimated the Republican vote. For example, Gallup overestimated the Republican vote by 6.8 percentage points in 1936 and by 5.4 percentage points in 1948. The difference was that in 1936 it did not matter.

A committee of the Social Science Research Council did an extended postmortem on the failure of the polls. Some conclusions in the preceding paragraph are from their report. The polls had ignored the dangers inherent in their own past errors. In fact, the election was so close that no one could have predicted the outcome with confidence. Dewey could have won by carrying three states that he lost by less than 1 percent of the vote. The pollsters should have foreseen the possibility of a close contest and should have taken measures to check on late shifts in voter intentions, the committee concluded. [10]

The 1948 fiasco sealed the doom of quota sampling. Besides the judgment element in deciding which factors are important, the interviewer had too much latitude in finding people to fill the quotas. Selection procedures that removed the possibility of bias in selecting respondents were instituted in polling. Typically, these involved the selection of blocks in cities and small areal units in the countryside, and the specification of how the interviewer was to move

Table 15-2. Gallup Poll Accuracy in Presidential Elections.

Year	Final survey	Percent for winner	Error on winning candidate
1936	55.7% Roosevelt	62.5% Roosevelt	−6.8
1940	52.0% Roosevelt	55.0% Roosevelt	−3.0
1944	51.5% Roosevelt	53.3% Roosevelt	−1.8
1948	44.5% Truman	49.9% Truman	−5.4
1952	51.0% Eisenhower	55.4% Eisenhower	−4.4
1956	59.5% Eisenhower	57.8% Eisenhower	+1.7
1960	51.0% Kennedy	50.1% Kennedy	+0.9
1964	64.0% Johnson	61.3% Johnson	+2.7
1968	43.0% Nixon	43.5% Nixon	−0.5
1972	62.0% Nixon	61.8% Nixon	+0.2
1976	48.0% Carter	50.0% Carter	−2.0
1980	47.0% Reagan	50.8% Reagan	−3.8
1984	59.0% Reagan	59.2% Reagan	−0.2

SOURCE: George Gallup. *The Sophisticated Poll Watcher's Guide* Princeton University Press, 1972, p. 215, and The Gallup Organization, Inc., Princeton, NJ.

through the area and select households. Areas were used as clusters and systematic (every nth) selection of household units encountered in a predesignated route became the standard.

Table 15-2 gives the record of Gallup Poll accuracy for presidential elections from 1936 to 1984. Final polls before elections are usually based on around 2500 persons. A random sample of 2500 has a two standard error range of ± 2 percentage points for a major candidate. It follows that Gallup was fortunate in 1960 that his point estimate for Kennedy was on the correct side of 50 percent.

The leading polls today include statements of sampling error along with their sample findings. In addition to putting readers on warning, it is a protection of their reputations in polling and commercial research. Most polling organizations make their steady living from commercial marketing and opinion research. Elmo Roper had this in mind in 1944 when he said:

Marketing research and public opinion research have demonstrated that they are accurate enough for all possible commercial and sociological purposes. We should protect from harm this infant science which performs so many socially useful functions, but which could be wrong in predicting elections . . . . [11]

Estimating Sample Size

Suppose a polling organization wants odds of 95 to 5 that the sample percentage voting for a major candidate will be within 3 percentage points of the final vote. Assuming that respondents who indicate they intend to vote actually do vote, how large a random sample will be required?

The statement about desired accuracy says that two standard errors should not exceed 3 percentage points, or that the standard error should be kept to 1.5 percentage points, or 0.015. The basic equation for the standard error of the proportion is

$$SE(p) = \sqrt{\frac{P(1 - P)}{n}}.$$

This relation, restated to solve for sample size, is

$$n = \frac{P(1 - P)}{[SE(p)]^2}.$$

Now, we do not know P, the true proportion voting for the candidate. But the product in the numerator cannot be larger than 0.25, and it will be close to this for major candidates in elections. A sample size that will take care of all contingencies is, then,

$$n = \frac{0.50(1 - 0.50)}{(0.015)^2} = 1111.$$

Let us redo this example using the coefficient of variation. If a population is split 50–50 in its votes for two major candidates (successes and failures or 1s and 0s), then the population mean is $0.50(P)$ and the population standard deviation is $\sqrt{0.50(1 - 0.50)} = 0.50$. The coefficient of variation is $100 * 0.50/0.50 = 100$ percent. The desire to estimate a more or less evenly divided population within 3.0 percentage points with 95 percent confidence amounts to a desire to limit two times the relative error to $(3.0/50) = 6$ percent with 95 to 5 odds. In these terms the formula for sample size is

$$n = (CV/RE)^2,$$

where CV stands for the coefficient of variation and RE stands for relative error. In our example,

$$n = (100/3.0)^2 = 1111$$

as before.

The point of having a formula for calculating a sample size that uses the coefficient of variation is that we have freed ourselves from units of particular variables. Figure 15-1 shows a number of general population shapes and the resulting coefficients of variation. The middle shape in the top panel is the one we were just considering—a Bernoulli population in which P is close to 0.50.

Many surveys, unlike election polls, are conducted to make estimates of a number of variables. For example, a national health survey might contain questions on incidence of a large number of different conditions, dollar expenditures on health insurance, distance to nearest emergency medical care

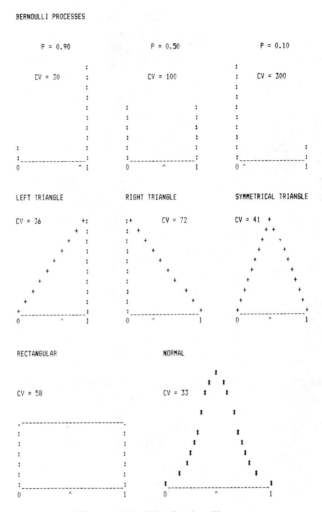

Figure 15-1. Distribution Shapes.

unit, intensive care unit, and so on. In short, we have a number of variables, some of which will be summarized in proportions and some of which will be summarized in means and standard deviations.

In planning surveys, the variables with large coefficients of variation will be the ones that will dictate sample size if the sponsors of the survey want the same relative error for many variables. As Figure 15-1 indicates, these will be Bernoulli populations with low proportions and right triangular distributions. Figure 15-1 is constructed for variables that have a low extreme value of zero, and is scaled so that the range of variation is 1.0. This applies naturally to Bernoulli populations. A normal distribution with this range will have a mean of 0.50. The standard deviation will be about one-sixth of the range. So the coefficient of variation will be $100(\frac{1}{6})/0.50 = 33$ percent.

A National Sample

We conclude our chapter on polls and surveys by describing a national sampling plan. The plan is for the national probability sample maintained by the Gallup Organization. Many other national sampling plans would have some of the same features.

First, the population of the United States is divided into seven size-of-community and urbanization groups and eight geographic regions. The result is $7 \times 8 = 56$ groups, or strata. The places within each strata are arrayed in a geographic order, and using census sources the population is cumulated through the array of places so that zones of equal population can be defined. Then, localities are selected within each zone with probability of selection proportional to population in the localities.

In the second stage of selection, the localities that were selected in the first stage are further subdivided and subareas selected with probabilities proportional to population size. In rural areas lacking small area population data, small geographic areas are defined and selected with equal probability. Within the selected subareas, individual blocks or block-areas are drawn with probability proportional to the number of dwelling units as reported in the Census of Housing and other sources. In rural areas lacking housing statistics, small areas are defined and selected with equal probability.

Within each third-stage area selected (block or area segment) a random starting point is selected on a map of the area. The interviewer starts at this point and follows a predetermined path, attempting to complete an interview at each household along the route. Interviewing is done on weekends and weekday evenings to maximize the chances of finding people at home.

About 300 second-stage sampling locations are included in the national sample. These are selected in a way that permits the division of the entire sample into a number of subsamples, each of which is an independent national sample. This feature is called *replicate subsampling*. The method was originated by P. C. Mahalanobis in India in 1936. W. Edwards Deming visited India in 1946 and 1951–1952, and was instrumental in spreading the method in the United States. [12] It allows standard errors of the survey to be easily calculated from the differences among the independent subsamples.

We see from this description that there are elements of stratification and elements of clustering in the design. Gallup reports that typical sampling errors for the design with a sample size of 1500 are around 3 percentage points (at 95 percent confidence) for percentages from 30 to 70. [13] The 95 percent error range for a random sample of 1500 is about ± 2.5. Thus, the design is about 20 percent less efficient than a random sample. That is, the *design effect* (DEFF) of the sample is about 1.2.

References

1. Morris Slonim, "Sampling in a Nutshell," *Journal of the American Statistical Association*, **52**, No. 2 (June 1957), 143–161.
2. Bill Williams, *A Sampler on Sampling*, Wiley, New York, 1978, pp. 5–10.
3. W. J. Youden, "Random Numbers Aren't Nonsense," *Industrial and Engineering Chemistry*, **49**, No. 10 (October 1957), 89A.
4. Warren Weaver, *Lady Luck*, Anchor Books, Garden City, NY, 1963, pp. 239–240.
5. J. A. Gulland, *Manual of Sampling and Statistical Methods for Fisheries Biology*, United Nations, New York, 1966.
6. W. G. Cochran, F. Mosteller, and J. W. Tukey, "Principles of Sampling," *Journal of the American Statistical Association*, **49**, No. 265 (March 1954), 13–35.
7. George Gallup, *The Sophisticated Pollwatcher's Guide*, Princeton University Press, Princeton, NJ, 1972.
8. Michael Wheeler, *Lies, Damn Lies, and Statistics*, Liveright, New York, 1976, Ch. 4.
9. Quotation from Wheeler, op. cit., p. 73.
10. F. Mosteller, *et. al. The Pre-election Polls of 1948*, Social Science Research Council, New York, Bulletin 60, 1949.
11. Wheeler, op. cit., p. 73.
12. W. Edwards Deming, *Sample Design in Business Research*, Wiley, New York, 1960, p. 186.
13. *Design of the Gallup Sample*, The Gallup Organization, Princeton, NJ, undated.

CHAPTER 16

Quality Control

In our chapter on the normal distribution we illustrated the important idea of the probability distribution of the sample mean with an example of a process control chart. While the probability distribution of the sample mean based on a known population standard deviation was known since the time of Gauss, the application to manufacturing control, as suggested in the example, did not take place until the third decade of the twentieth century. It appears to have been an outgrowth of the emphasis on small samples and exact sampling distribution theory begun by Gosset in 1908 and extended greatly by R. A. Fisher by the early 1920s. [1]

Quality Control to 1930

Attempts to control the outcomes of fabricated goods existed before the industrial revolution. The Bible says that God gave Noah the specifications for the Ark. The guilds of medieval times controlled the quality of output in their crafts by systems of apprenticeship. Quality control rested then in reliance on the artisan's training and reputation.

An interesting example of a sampling inspection scheme for a continuous process is the sampling of coinage from the Royal Mint in London, a practice that dates back over eight centuries. [2] The process of the "trial of the pyx" was originally operated as a check on the minting of coins by a minter operating under contract to the Crown. If the minter put too much of the king's gold in the coins, persons coming into possession of the coins could melt them down and sell them back to the mint at a profit to themselves (and a loss to the Crown). If the minter put too little gold into the coins, the currency would be debased and the profit from the underage would go to the minter.

The control scheme involved having a section of a trial plate of the king's gold stored away as a standard. Then, one coin from each day's production would be placed in a box (called a pyx). On a frequency varying from yearly to every few years a "trial of the pyx" would be declared. The box would be opened and the coins would be collected, counted, weighed, and assayed. A tolerance above and below the standard set by the trial plate was allowed. Any shortage beyond the tolerance was used to estimate an amount to be paid back to the king on coinage minted since the previous trial.

Mass production requires interchangeability of parts which, in turn, requires adequate production controls. In 1798 Eli Whitney, the inventor of the cotton gin, secured a contract from Congress to manufacture 10,000 army muskets. He intended to mass produce interchangeable parts, but the attempt failed because parts that were supposed to be "the same" varied too much to be truly interchangeable. Whitney's ideas were eventually validated when better gauges were developed to guide production in the early 1800s.

While informal methods of sampling output to assure that products met a specification doubtless existed, methods that were grounded in probability and statistical principles were rarely used in industry before the 1920s. The basic concept of the contol chart was unknown until that time. [3] Some early applications of statistics were made by persons at the Bell Telephone System who were using the normal and Poisson distributions in connection with telephone traffic problems before 1910. Recall that it was about this time that "Student" was hired by the Guinness Brewery in London—one of the first statisticians to specialize in industrial applications.

In 1924 a group was established at the Western Electric Company to look into the control of quality of manufactured products. One of the first members of the group was Walter Shewhart (1891–1967). Shewhart developed the principles of the control chart in a matter of days, and these were circulated in company memoranda and a few published papers from 1924 to 1926. In the succeeding 5 years the ideas were tested out within the Bell Telephone System and practical procedures developed for integrated quality control programs.

The two related aspects of statistical quality control are process control (in which the control chart is the primary tool) and sampling inspection. While we have had a brief exposure to the control chart, we have not illustrated the ideas in sampling inspection. We need to do this at this point.

Sampling Inspection

We will illustrate some of the principles of sampling inspection with an example of an outgoing quality check. A manufacturer produces lots of 5000 fuses each. To maintain the quality of outgoing shipments a quality control person has suggested the following.

Decision rule: From each lot select a sample of 20 fuses at random. If all the fuses operate properly ship out the lot. If one or more fuses are defective, then inspect the entire lot and replace any defective fuses with ones that operate properly.

The consequences of this decision rule can be readily evaluated. We could find from the binomial distribution the probability of shipping out a lot of any specified proportion defective, given that such a lot was submitted for inspection. Under the suggested rule this is the probability of no occurrences in 20 trials of an event whose probability is equal to the proportion of defective fuses in the lot.

In Table 16-1 we show these probabilities for the rule suggested, and for a second rule. The second rule requires a sample of 50 fuses with the lot to be accepted if less than 2 defective fuses out of the 50 are found. We see that under the first sampling scheme the probability of shipping a 1 percent defective lot is about 0.82, while the probability of shipping a 10 percent defective lot is about 0.12. Assume that the state-of-the-art is such that a 1 percent defective lot is a good lot and a 10 percent defective lot is a bad lot. The probability of *not* shipping a 1 percent defective lot, $1 - 0.82 = 0.18$ is called the *producer's risk*, and the probability of shipping a 10 percent defective lot is termed the *consumer's risk*. The table shows risks at other levels of lot quality. If 2 percent defective is a fairly good lot then the probability of not shipping out such a lot is a producer's risk, and if 9 percent is a fairly bad lot, then the probability of shipping out such a lot is the consumer's risk. The point is that at some low percentage defective we are concerned that the lot not fail inspection, and at some high percentage defective we are concerned that the lot not pass inspection. In between there is an area of indifference. The entire

Table 16-1. Probability of Accepting Lots of Differing Proportions Defective Under Two Sampling Schemes.

Lot proportion defective	Sample of 20 Accept if 0 defects	Sample of 50 Accept if 0 or 1 defect
0.00	1.000	1.000
0.01	0.818	0.911
0.02	0.668	0.736
0.03	0.544	0.555
0.04	0.442	0.400
0.05	0.358	0.279
0.06	0.290	0.190
0.07	0.234	0.126
0.08	0.189	0.083
0.09	0.152	0.053
0.10	0.122	0.034

relation of the probability of accepting the lot to the lot quality is called the *operating characteristic* curve for the inspection scheme.

Looking at the operating characteristic curve for the second inspection scheme we find that both the producer and consumer risks are reduced. The probability of rejecting a 1 percent defective lot is only $1 - 0.911 = 0.089$, and the probability of shipping out a 10 percent defective lot has been reduced to 0.034. We should expect that with a larger sample there is a rule that reduces both kinds of errors. Of course, the inspection costs are increased with the larger sample size.

We might ask what will be the quality of the lots shipped out under each inspection scheme. These are not difficult to calculate on a conditional basis. For example, under the second scheme we just saw that the expectation is that 8.9 percent of all 1 percent defective lots will be subject to complete inspection to replace defective fuses with good ones. For this fraction of lots there will be 0.0 percent defective fuses, and for 91.1 percent of the 1 percent defective lots there will be 1 percent defectives in the lots shipped out. Thus the *average outgoing quality*, or AOQ, for 1 percent defective lots submitted to the inspection process will be

$$\text{AOQ (1\% incoming)} = 0.089(0) + 0.911(1.0) = 0.0911\%.$$

For incoming (to inspection) 4 percent defective lots under the second scheme the AOQ will be

$$\text{AOQ (4\% incoming)} = 0.600(0) + 0.400(4.0) = 2.400\%,$$

because the probability of passing a 4 percent defective lot is 0.400. The remaining 60 percent of 4 percent defective lots will be subject to complete inspection and replacement of any defectives.

In Figure 16-1 the operating characteristic curves for the two schemes are plotted as well as the AOQ curves. We have already talked about the operating characteristic, or OC curves. The AOQ curves are low at both quality extremes and reach a maximum at some level of intermediate quality where roughly half of the lots are 100 percent inspected and the rest passed along. The important point about AOQ is that the inspection scheme can guarantee that the average outgoing quality (AOQ) will not exceed that limit (AOQL). Exactly what quality will be delivered will depend on the quality mix of lots incoming to inspection, but the limit can be guaranteed. Of course, if the quality submitted to inspection is poor, the inspection and replacement of defective items will be costly. To avoid this is where process control with the aid of Shewhart control charts fits in.

The 1930s

Harold F. Dodge and Harry G. Romig joined Western Electric shortly after Walter Shewhart. Their contribution was the development of sampling in-

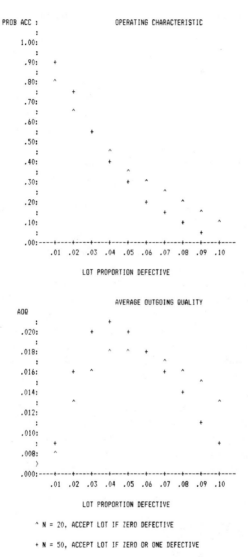

Figure 16-1. Operating Characteristic and Average Outgoing Quality for Two Sampling Plans.

spection plans. The basic features of sampling inspection plans were in operation in the Bell Telephone System by the time Shewhart summarized the state of statistical quality control in his 1931 book *Economic Control of Quality of Manufactured Products* (Van Nostrand).

In the 1930s the gospel of statistical quality control began to be spread slowly. Technical journals carried articles on the subject. Shewhart had given a course at the Stevens Institute of Technology based on notes for his book, and in 1932 a quality control course was begun in Industrial Engineering at Columbia University. Leslie E. Simon began quality control work for the U.S. Army at the Picatinny Arsenal in 1934. By the late 1930s Shewhart, Simon, and Dodge had been instrumental in having the Army Ordnance Department develop war standards for quality control. [4]

Shewhart conducted a course at the U.S. Department of Agriculture in 1938, and the same year a conference and lectures were held at the Massachusetts Institute of Technology. L. H. C. Tippett of the British Cotton Industry Research Association participated in the conference. At the time, however, H. A. Freeman of MIT could find only a dozen industrial firms in the United States that had introduced statistical quality control in a thorough way. [5]

L. H. C. Tippett had begun work in 1925 for the British Cotton Industry Research Association after studying under Karl Pearson on a scholarship from the association. He mentions also studying for a time with R. A. Fisher. Researchers had found uncontrolled variation in fiber, yarn, and fabric properties and decided that statistics might be of help to them. [6] Tippett was an early user of both the experimental methods of Fisher and of quality control in British industry. Another early advocate for industrial applications of statistics in Britain was Bernard P. Dudding of the General Electric Company. Dudding mentions that studies aimed at reducing variability in efficiency of lamps made at General Electric in the mid-1920s employed analysis of variance. By 1930 Shewhart control charts had been extended to regular operations in about 20 stages of glass and filament manufacture and in lamp assembly. [7]

The use of statistical quality control in Britain during the 1930s spread more rapidly than in the United States. Shewhart had visited London in 1932, and E. S. Pearson produced an early text on industrial quality control. By 1937 applications had been made in coal, coke, cotton yarns and textiles, woolen goods, glass, lamps, building materials, and chemicals. [8]

QC Goes to War

The outbreak of World War II in 1939 turned the United States toward expansion of its own military forces in the interests of preparedness. Cooperation between the Bell Telephone Labs and the Army Ordnance Department had been established earlier. After Pearl Harbor, General Leslie E. Simon, who had been guided by Shewhart to start his work at Picatinny Arsenal in 1934, formed a group consisting mainly of Dodge and Romig and George W. Edwards of Bell Laboratories and G. R. Gause of Army Ordnance to begin the development of sampling inspection tables for ordnance procurement. These tables were put into operation in 1942, and at the same time three-day training conferences were instituted for ordnance inspectors.

Early in 1942, W. Allen Wallis, on behalf of the Committee on Instruction in Statistics at Stanford University, wrote to W. Edwards Deming (then of the U.S. Bureau of the Census) for advice on how a contribution to the war effort might be made. Deming responded with four single-spaced pages on Chief of Ordnance, War Department letterhead suggesting quality control courses for executives, engineers, and inspectors in industry. A ten-day course was designed by E. L. Grant and Holbrook Working of Stanford and given at Stanford in July. The course was reduced to eight days and given in Los Angeles in September 1942. [9]

In February 1943 Walter Shewhart and S. S. Wilks of Princeton University went to Washington representing the Committee on Applied Mathematics of the National Research Council. They were looking for a way to stimulate the use of statistical quality control in the war industries. In England a government consulting staff had been set up to provide advice to industry, and this was the only precedent available. They met with Paul T. Norton, Jr. of the Office of Production Research and Development, who had just come from Virginia Polytechnic University to head the Division of Industrial Processes and Products of OPRD. Norton was a college room-mate of E. L. Grant, who was in Washington for a regional meeting of the Emergency Science and Management War Training Program (ESMWT), sponsor of the Stanford quality control training courses. [10]

While Norton had heard of quality control, he was not convinced that OPRD could take any actions that would contribute to the wartime emergency. Norton happened to have dinner with Grant, and told him of the visit from Shewhart and Wilks. Grant was able to report on applications made by the Ontario Works of the General Electric Company in California that resulted from the Stanford course given in Los Angeles. Grant put Norton in touch with Deming, and a general plan for an expanded program of quality control training in all regions of the country was put together by Deming, Norton, Leslie E. Simon, and Dean George Case of Stanford. Originally directed by Holbrook Working of Stanford and later by Edwin G. Olds of Carnegie Institute of Technology, the OPRD program eventually enlisted over 10,000 persons from 810 organizations in 40–80 hour intensive quality control courses given at 43 different educational institutions in the country from 1943 to 1945. These courses provided the nucleus for establishment of quality control societies in the different locations, and after the war these organizations joined together to form the American Society for Quality Control. George D. Edwards of the Bell Telephone Laboratories was the first president of ASQC and Walter Shewhart was its first honorary member. [11]

In 1942 a Statistical Research Group (SRG) was formed at Columbia University, supported by the Office of Scientific Research and Development, to do statistical work for the armed services and certain suppliers. The group was organized by Harold Hotelling of Columbia University and directed by W. Allen Wallis. The members drawn to this group reads like a "Who's Who" in statistics 20 years later. They included (alphabetically) Kenneth J. Arnold,

Rolling F. Bennett, Julian H. Bigelow, Albert H. Bowker, Churchill Eisenhart, H. A. Freeman, Milton Friedman, M. A. Girschik, Millard W. Hastay, Frederick Mosteller, Edward Paulson, L. J. Savage, Herbert Solomon, George J. Stigler, Abraham Wald, and Jack Wolfowitz. Warren Weaver of the Applied Mathematics Panel of the National Defense Research Committee of the Office of Scientific Research and Development also strongly influenced and guided SRG.

The group worked on practical and immediate problems such as the mix of ammunition for fighter planes, optimal settings for proximity fuses for artillery shells under various combat conditions, the comparative effectiveness of different armament on fighter aircraft, and optimal angles for aircraft torpedo attacks on warships. In addition they gave consultative and informal ongoing assistance to the Army, Navy, and NDRC. [12]

The most significant contribution of SRG was the development and application of *sequential analysis*. A navy captain had raised with Wallis the question of how large a sample size was needed for comparing two percentages to a desired degree of precision in testing alternative methods in ordnance fire. Wallis and Paulson worked up some answers, which ran to many thousand rounds for the desired precision. The captain commented that a seasoned ordnance officer might see after a few thousand, or even a few hundred rounds that the experiment need not be completed because of the obvious superiority of one of the methods. But you could not give any leeway to testing personnel because they had no such judgment. Wasn't there some rule that could be developed that would tell when to terminate an experiment earlier than planned?

Wallis discussed the problem with Milton Friedman who recognized that a rigorous solution to the problem would represent a real contribution to statistical theory. After discussing whether their mathematical powers were up to the task, they decided to call for help. They turned first to Wolfowitz, who then brought Abraham Wald into the discussion along with Hotelling. Wald took on the problem, and in a short time worked out the basic approach to sequential analysis. H. A. Freeman undertook to translate Wald's work into inspection applications. The first large scale application was made with Freeman's assistance in the Quartermaster Inspection Service of the U.S. Army. [13].

Wald continued to make pioneering contributions to statistical decision theory until his untimely and tragic death in an airplane accident in 1950. Wallis and Bowker went on to careers in statistics and chancellorships of the University of Rochester and the University of California. Milton Friedman achieved fame in economics at Chicago and received the Nobel Prize in economics in 1976. Mosteller and Eisenhart have enjoyed illustrious careers at Harvard and at the National Bureau of Standards. Bowker, Eisenhart, Mosteller, and Wallis have been president of the American Statistical Association, and Friedman and Stigler have been president of the American

Economic Association. Eight members of the SRG have been president of the Institute of Mathematical Statistics. [14]

During the war single sampling plans of the kind illustrated earlier, double sampling plans, and sequential sampling plans were put into operation by the services. For example, a manual prepared by the Statistical Research Group for the Office of Procurement and Material of the Navy Department contained the operating characteristic curves and average outgoing quality limits for a number of single, double, and sequential sampling plans. [15] Here is an illustration of a double sampling plan.

> Double Sampling: Take a sample of 100 from a lot. If 2 or fewer defectives are found, accept the lot. If more than 6 defectives are found, reject the lot. If between 3 and 6 defectives are found, take a further sample of 100 and if the combined number of defectives is 6 or less accept the lot. If the combined number of defectives exceeds 6, reject the lot.

The operating characteristic curve for this sampling plan is shown in Figure 16-2. [16] The advantage of double sampling plans is that one can achieve the

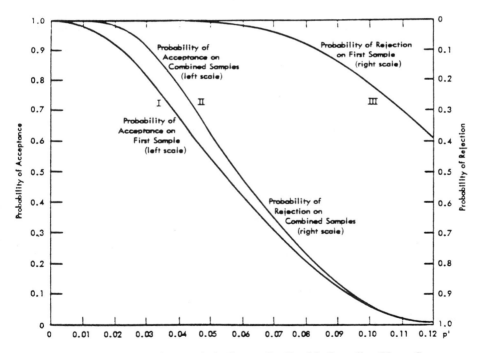

Figure 16-2. Operating Characteristic Curves for Double Sampling Plans. SOURCE: Duncan, A. J., *Quality Control and Industrial Statistics*, p. 180. Reproduced by permission.

same operating characteristic curve as a single sampling scheme but with a smaller average sample size, and thus a smaller inspection cost. By the same operating characteristic is meant a desired limitation on both producer and consumer risks. For example, a specification might be that the sampling, whether single or double, limit the risk of having a 1 percent defective lot rejected to 0.05 at the same time that the risk of accepting an 8 percent defective lot is limited to 0.10. Quality control manuals provide the sampling schemes to meet such specifications.

Sequential Sampling

Rather than having a fixed sample size or a fixed pair of sample sizes, sequential sampling uses preset limits for the number of defectives that will result in accepting a lot and the number of defectives that will result in rejecting a lot *at any stage* as sampling proceeds. As long as the (cumulative) number of defectives has not reached either limit, sampling continues. Figure 16-3 shows a sequential sampling chart for a sequential sampling plan that meets the specifications for producer and consumer risk mentioned above. Wald and his associates developed the formulas that provide the sequential acceptance and rejection limits. They also showed that at the critical lot quality levels no other sampling plan can have a smaller expected sample size.

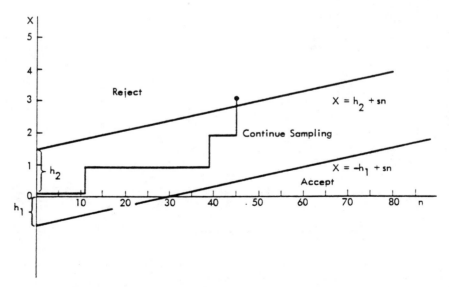

Figure 16.3. Sequential Sampling Chart. SOURCE: Duncan, A. J., *Quality Control and Industrial Statistics*, p. 192. Reproduced by permission.

In general, sequential sampling plans reduce the expected sample sizes to about half the number required by single sample plans. [17]

Post-World War and Today

After World War II Japanese industrial leaders realized that to compete in international markets would require a revolutionary change in the quality of Japanese products. Industrial leaders invited W. Edwards Deming to give an eight-day lecture course to 230 engineers of leading Japanese manufacturing companies in 1950. [18] This course was instrumental in what was to develop into a massive training program that included top executives, middle management, and foremen in addition to engineers. J. M. Juran began in 1954 to provide training in Japan on the managerial aspects of quality control. [19]

By 1962 training in quality control in Japan had spread to production workers. Training was given to small groups of workers in the same department, usually seated around a table. These quality control *circles* involved workers in selection, analysis, and solution of problems related to product quality. The culture of mutual worklife responsibility between worker and company created a climate in which quality circles were immensely successful. In 1967 J. M. Juran predicted that "the Japanese are headed for world quality leadership, and will attain it in the next two decades, because no one else is moving there at the same pace." [20]

Juran draws a contrast between the Japanese quality revolution and earlier industrial development in the United States that followed the management theories of Frederick W. Taylor. Taylor's system separated planning from execution on the grounds that first-line supervisors and workers did not have the education needed for tasks such as establishing work methods, training employees, and setting standards. In the early years of the twentieth century spectacular increases in productivity resulted from application of Taylor's system, but the effects on product quality could be questioned. Quality control activity in the United States began in independent inspection departments set up to prevent quality deterioration under Taylor-like systems. [21]

There are other aspects of the Japanese success story in quality. These include more complete removal of product defects in the development and design stage, acceptance of longer payback periods for recovering new product investments, and novel market structures for servicing and feeding-back field service data, and closer cooperation with vendors. In the manufacture of television sets, Japanese management techniques are credited with cutting the levels of defects by 40 times, and in some cases by a factor of 100. [22] These magnitudes have been confirmed in a recent study by David A. Garvin which placed the comparison for assembly line defect rates at 70 times. [23]

These comparisons with Japanese achievements have sparked new concern with product quality in the United States. Deming is in demand as a consultant to U.S. auto makers. [24] One success-story is told of cutting initial defects in Pontiac's 2.5-liter engine from 40 percent to 4 percent. [25]

References

1. See Acheson J. Duncan, *Quality Control and Industrial Statistics*, 4th edn, Irwin, Homewood, IL, 1974, Ch. 1 for a general historical review of quality control.
2. Stephen M. Stigler, "Eight Centuries of Sampling Inspection: The Trial of the Pyx," *Journal of the American Statistical Association*, **72**, No. 359 (September 1977), 493–500.
3. S. B. Littauer, "The Development of Statistical Quality Control in the United States," *The American Statistician*, **4** (December 1950), 14–20.
4. Ibid., p. 17.
5. Duncan, op. cit., p. 2.
6. L. H. C. Tippett, "The Making of an Industrial Statistician," in *The Making of Statisticians* (J. Cani, ed.), Springer-Verlag, New York, 1982, pp. 182–187.
7. Bernard P. P. Dudding, "The Introduction of Statistical Methods o to Industry," *Applied Statistics*, **1** (1952), 3–20.
8. Duncan, op. cit., p. 6.
9. Holbrook Working, "Statistical Quality Control in War Production," *Journal of the American Statistical Association*, **40**, No. 232 (December 1945), 425–447.
10. Eugene L. Grant, "Industrialists and Professors in Quality Control," *Industrial Quality Control*, **X**, No. 1 (July 1953), 31–35.
11. See references by Working op. cit., Duncan op. cit., Littauer op. cit., Grant op. cit.
12. W. Allen Wallis, "The Statistical Research Group, 1942–1945," *Journal of the American Statistical Association*, **75**, No. 370 (June 1980), 320–330.
13. Ibid., p. 326.
14. Ibid., p. 324.
15. "Acceptance Sampling, "American Statistical Association, Washington, DC, 1950, p. 34.
16. From Duncan, op. cit., p. 180.
17. Duncan, op. cit., p. 195.
18. W. Edwards Deming, *Elementary Principles of the Statistical Control of Quality*, Nippon Kagaku Gijutsu Remmei, Tokyo, 1950.
19. Frank M. Gryna, Jr., *Quality Circles: A Team Approach to Problem Solving*, American Management Association, New York, 1981.
20. J. M. Juran, "Japanese and Western Quality—A Contrast," *Quality Progress*, **XI** (December 1978), 18.
21. Ibid., p. 11.
22. Ibid., p. 14.
23. David A. Garvin, "Quality on the Line," *Harvard Business Review*, **61** (September–October 1983), 65–75.
24. Jeremy Main, "The Curmudgeon who Talks Tough on Quality," *Fortune* (June 25, 1984), 118–122.
25. "Dr. Deming Shows Pontiac the Way," *Fortune* (April 18, 1983), 66.

CHAPTER 17

Principal Components and Factor Analysis

In Chapter 10 on correlation and regression we used an example of the numbers of runs scored in the 1975 and 1976 seasons by American League baseball clubs. The data were shown in Table 10-1 and the correlation was 0.63. Figure 17-1 shows the data plotted as a scatter diagram in standardized units.

Also shown in the figure are a series of concentric circles. Each circle represents the locus of equal values of the sum of squared standard scores for the two variables, that is

$$\text{chi-square} = Z^2(X) + Z^2(Y).$$

We will call these circles chi-square isoquants. Geometrically, the chi-square value for any point is the square of the hypotenuse formed by the right triangle whose base is the X-distance from the origin (0, 0) and whose height is the Y-distance from the origin. The Z-scores and chi-square values for the data are given in Table 17-1. Since the square roots of the chi-square values are the distances of each point from the center point of the data, they measure the variation of the points in the two-variable space. We see that the Tigers, Angels, and Red Sox are furthest from the center, while the Brewers, Orioles, and Rangers and closest to the center. The mean chi-square value is 2.0.

Recall that the correlation of X and Y is the average cross-product of the standard scores. Plotted on the standardized scatter diagram, the correlation of Y with respect to X *would show* a slope of 0.63 standard deviations change in Y per standard deviation change in X. The correlation of X with respect to Y would show graphically the change of 0.63 standard deviations change in X per standard deviation change in Y. These lines are not the same, *but would form a positively* sloping "X," crossing at the origin. The angle formed by $r(YX)$ would be about 32 degrees from the X-axis, while that formed by $r(XY)$

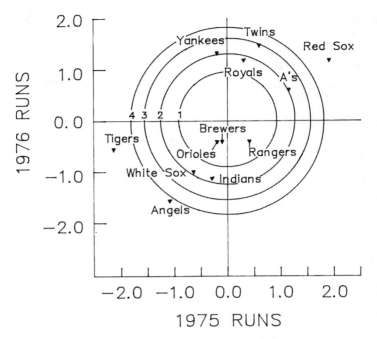

Figure 17-1. Chi-Square Isoquants for Baseball Runs Data.

Table 17-1. Z-Scores and Chi-Square Values
for Runs Data.

Team	1975 Runs Z(X)	1976 Runs Z(Y)	Chi-square
Yankees	1.30	−0.16	1.72
Orioles	−0.42	−0.14	0.20
Red Sox	1.08	1.89	4.73
Brewers	−0.48	−0.04	0.23
Indians	−1.18	−0.27	1.46
Tigers	−0.57	−2.14	4.90
A's	0.62	1.21	1.85
Twins	1.50	0.60	2.62
White Sox	−0.93	−0.62	1.26
Rangers	−0.47	0.43	0.40
Royals	1.04	0.35	1.20
Angels	−1.49	−1.11	3.44
Average	0.00	0.00	2.00

is 32 degrees from the Y-axis, leaving the angle between the two regression lines at about 26 degrees.

The Principal Component

The Cartesian coordinates of the scatter diagram of Figure 17-1 are oriented as we are accustomed to seeing them. Imagine now that the coordinates are rotated counterclockwise until the X-axis is oriented along the scatter of the points from "southwest" to "northeast." We will specify the meaning of this more exactly in a moment. The result is the reoriented axes in Figure 17-2. The chi-square isoquants now become ellipses with their major (longer) axis oriented to the scatter of the data. The chi-square values for the data points are unchanged. They are still the square of the distance of each point from the origin. The distance now forms the hypotenuse of a different right triangle. The base of the right triangle is the distance of the data point measured along the major axis of the ellipse, and the height of the triangle is the distance of the point measured on the minor axis. This relationship is shown in Figure 17-3.

The ellipses in our figures are lines of equal values of chi-square, the sum of the squared values for the two coordinate axes of measurement. Bivariate data

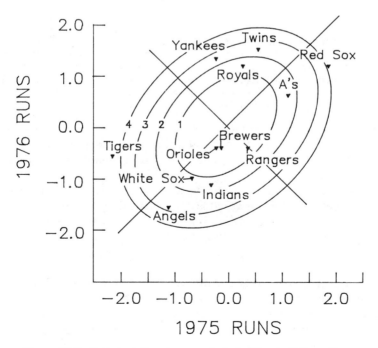

Figure 17-2. Principal Components Axis for Baseball Runs Data.

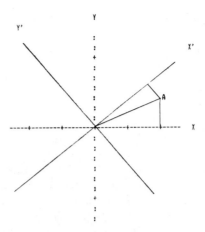

Figure 17-3. Axis Rotation.

are often distributed so that the lines of equal frequency density will follow the elliptical shapes that we see in Figure 17-2. This was recognized by Galton in studying some data on heights of parents and their offspring as adults. He used a correlation chart in which he divided the ranges of X and Y into intervals. The chart then forms a number of cells. Galton had written at intersecting lines the average frequencies in the adjoining four cells. He wrote

> But I could not see my way to express the results of the complete table in a single formula. At length, one morning, while waiting at a roadside station near Ramsgate for a train, and poring over the diagram in my notebook, it struck me that the lines of equal frequency ran in concentric ellipses. The cases were too few for certainty, but my eye, being accustomed to such things, satisfied me that I was approaching the solution. [1]

Galton then sent his problem, disguised as a problem in mechanics, to J. Hamilton Dickson, a mathematics tutor at Cambridge. Dickson sent him the formal solution, commenting that "most of his mathematical class were able to solve it." [2]

The original Cartesian coordinate axes are rotated until there is zero correlation between the two variables as measured against the rotated axes. The equation for an ellipse with a given value of chi-square, based on standardized scores on the rotated axes (indicated by the primes) is

$$\text{chi-square} = Z^2(X') * (1 + r) + Z(Y') * (1 - r).$$

Suppose that X and Y are standardized with unit variance. The standard deviation of the distances along the major axis is the square root of $1 + r$, and the standard deviation of the distances on the minor axis is the square root of $1 - r$. We now have the variances (in terms of expectations, or averages)

$$E(\text{VAR}) = 1.63 + 0.37 = 2.00.$$

When measured on the original Cartesian axes, the average variances were

$$E(\text{VAR}) = 1.00 + 1.00 = 2.00.$$

What the rotation has done is to add the variation that the two variables (1975 and 1976 runs) have in common to form the principal component and to take it away to form the minor component.

The major axis is called the *principal component*. The formal development of principal components analysis is attributed to Harold Hotelling (1895–1973), who published the fundamental article in 1933. [3] Karl Pearson had used the same mathematical principle some 30 years earlier in an article entitled "On Lines and Planes of Closest Fit to Systems of Points in Space," but it is believed that Hotelling was unaware of Pearson's work in this regard. [4]

We have seen that our principal component accounts for 1.63 out of the total bivariate standardized variance of 2.00. It includes the elements of variation that 1975 and 1976 runs have in common. The remaining variance of $2.0 - 1.63 = 0.37$ is the standardized variance not shared between the two years. The Z-scores for the principle component are given in Table 17-2. We can note that the principal component ranks the teams very much like the average runs for the two years, which can be calculated from Table 10-1. The only difference is in the very last places where the Angels and Tigers are very close by either accounting.

The use of principal components with two variables has served to introduce the idea of rotating an axis of measurement and reflecting the values for points in the measurement space onto the rotated axis. It is a foregone conclusion that the principal component in the two-variable case will be an equally weighted linear combination of the two variables. The standardized values of the principal component will be

$$Z(C) = \frac{1+r}{2} * Z(X) + \frac{1+r}{2} * Z(Y).$$

It is not surprising that the component scores gives results similar to the mean runs by teams in this case.

Table 17-2. Z-Scores for Principal Component, 1975 and 1976 Runs.

Team	Z-Score	Team	Z-Score
Red Sox	1.65	Brewers	−0.29
Twins	1.17	Orioles	−0.31
A's	1.01	Indians	−0.80
Royals	0.77	White Sox	−0.86
Yankees	0.63	Angels	−1.44
Rangers	−0.02	Tigers	−1.50

For an example with more variables we decided to expand our runs data and deal with the total runs for 24 Major League baseball teams for the 5 years 1972 through 1976. If we had just expanded to 3 years our geometric reference would be a square box, with length, width, and height representing the variables V, X, and Y. These would be measured in standard deviation units from the three-dimensional center of the box. Instead of four sectors as we have for two dimensions (NE, SE, SW, and NW) we would have eight sectors in the box—four lower quadrants and four upper quadrants. If there were no correlations among the three variables, the data points would form a spherical pattern in this three-dimensional space. If there are intercorrelations among the variables, the scatter in the three dimensions will become elliptical. The long axis is like a rod extending from the center to the long ends of a football. The remaining axes are perpendicular to the long axis, and together form the two-dimensional section through the middle of the football. But this section is not necessarily circular (like the football). It can be elliptical.

Finding the principal components in the three-variable case amounts to rotating the original three-dimensional measurement axis until one axis is oriented as the rod through the length of the football. Then the remaining two axes in the plane through the center of the football are rotated together (so they stay at right angles) until one axis is oriented along the longer axis of that ellipse and the other along the shorter axis. In this new measurement space, the variables (V', X', and Y') will be uncorrelated. The principal component will represent the variation common to the three variables. Figure 17-4 shows the elliptical shape in three dimensions. [5]

The baseball runs data has five variables. The original data and the matrix of correlations between the variables are given in Table 17-3. We see that the runs scored in a season are correlated with the runs scored in other seasons. In fact, for the most part the runs scored in a season are more highly correlated with runs scored in the preceding or following year than with runs in seasons removed by more than a year. This is probably all to be expected.

Principal components analysis rotates the five measurement axes into a new set of orthogonal (right-angle) dimensions in which the variables are uncorrelated. The first dimension, or principal component, will be the weighted sum (or linear combination) of the original variables that explains the maximum multidimensional variance. With five variables, the sum of standardized variances is 5.0.

The remaining components explain, successively, the maximum of the variance not accounted for by previous components. With five original variables, principal components analysis will extract five components which together account for the original standardized variance of 5.0. It may be that only the first, or principal component, is of interest, however.

We will indicate the variables in the baseball runs data by the season. The principal component explains 2.89 (or 57.8 percent) of the total variance of 5.0. The next component explains 0.81, (or 16.2 percent), and the remaining components even less. An equation for the principal component is

$$C = 0.8509\ (75) + 0.8243\ (76) + 0.7867\ (74) + 0.6539\ (73) + 0.6628\ (72).$$

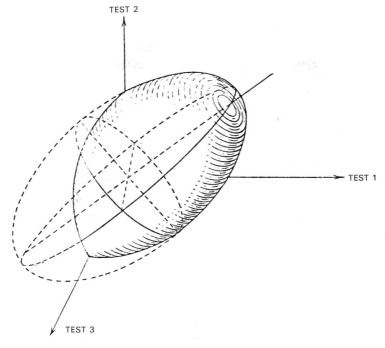

Figure 17-4. Three-Dimensional Ellipse.

Here, the coefficients are to be multiplied by the Z-scores for the variables. The principal component scores (C) represent the elements of variation that the different seasons have in common. The different years are not equally weighted in the principal component scores.

The coefficients are the correlations of the variables with the common factor, or principal component. The correlation of 1975 runs with the common factor is higher than, for example, 1972 runs. In factor analysis language, these are called *factor loadings*. One would say that 1975 runs are loaded more highly with whatever the seasons have in common than are 1972 runs.

The variance explained by a component is the sum of squares of the factor loadings for the variables on that factor. In Table 17-4 we show the factor loadings for all the components and then the squared factor loadings. There is no substantive interest here in components beyond the first one. But the table of squares of the factor loadings illustrates how variance is reassigned from the original variables to the components. Adding across, we see the unit standardized variances for the original variables. Adding down, we see how the variance is allocated to the different components. Both, of course, account for the total standardized variance of 5.0.

Again we find that the ranking of teams based on principal component scores is close to the ranking based on average runs for the five seasons. The results are shown in Table 17-5.

Table 17-3. Runs by Season for Major League
Teams and Correlation Matrix.

	1972	1973	1974	1975	1976
Phillies	503	642	676	735	770
Pirates	691	704	751	712	708
Cubs	685	614	669	712	611
Cards	568	643	677	662	629
Mets	528	608	572	646	615
Expos	513	668	662	601	531
Dodgers	584	675	798	648	608
Reds	707	741	776	840	857
Astros	708	681	653	664	625
Giants	662	739	634	659	595
Braves	628	799	661	583	620
Padres	557	641	671	681	730
Yankees	519	754	659	682	619
Orioles	640	738	696	796	716
Red Sox	493	708	647	675	570
Brewers	472	680	662	688	615
Indians	558	642	620	570	609
Tigers	604	758	689	758	686
A's	537	738	673	724	743
Twins	566	652	684	655	586
White Sox	461	619	690	714	616
Rangers	580	755	667	710	713
Royals	454	629	618	627	550
Angels	488	548	541	552	570

Correlation Matrix

	1972	1973	1974	1975	1976
1972	1	0.4145	0.4418	0.3642	0.3967
1973	0.4145	1	0.4219	0.3961	0.3681
1974	0.4418	0.4219	1	0.5969	0.5033
1975	0.3642	0.3961	0.5969	1	0.7743
1976	0.3966	0.3681	0.5033	0.7743	1

General Intelligence

Factor analysis refers to a family of techniques, of which principal components is a member. The pioneering paper in factor analysis, entitled "General
Intelligence" was written by Charles E. Spearman in 1904. The origins of
the ideas that Spearman took up can be found in Galton's approach to two
different problems, however. These are the measurement of inherited characteristics and the identification of criminals.

Galton concerned himself with a system in which two measures are each the
result of peculiar variation plus variation arising from a common source. The

Table 17-4. Factor Loadings and Squares of Factor Loadings for Baseball Runs Data.

Variable	Factor loadings Component				
	1	2	3	4	5
1975	0.8499	0.3949	0.0534	0.0706	−0.3375
1976	0.8234	0.3894	−0.0215	0.2905	0.2924
1974	0.7878	0.0087	−0.0573	−0.6066	0.0875
1973	0.6583	−0.4987	0.5546	0.1013	0.0094
1972	0.6635	−0.5046	−0.5240	0.1688	−0.0462

	Squared factor loadings					Total
1975	0.7223	0.1559	0.0029	0.0050	0.1139	1.000
1976	0.6780	0.1516	0.0005	0.0844	0.0855	1.000
1974	0.6206	0.0001	0.0033	0.3680	0.0077	1.000
1973	0.4334	0.2487	0.3076	0.0103	0.0001	1.000
1972	0.4402	0.2546	0.2746	0.0285	0.0021	1.000
Total	2.8945	0.8109	0.5889	0.4962	0.2093	5.000

Table 17-5. Standardized Principal Component Scores and Average Runs per Season for Major League Baseball Clubs, 1972–1976.

Team	Z-Score	Average	Team	Z-Score	Average
Reds	2.67	784	Giants	0.03	658
Orioles	1.33	717	Yankees	−0.02	647
Pirates	1.23	713	Cards	−0.25	636
Tigers	0.99	699	Twins	−0.38	629
A's	0.70	683	White Sox	−0.41	620
Rangers	0.68	685	Brewers	−0.41	623
Phillies	0.40	665	Red Sox	−0.54	619
Dodgers	0.39	663	Expos	−1.03	595
Astros	0.18	662	Indians	−1.06	600
Padres	0.14	656	Mets	−1.16	594
Cubs	0.10	658	Royals	−1.38	576
Braves	0.05	658	Angels	−2.24	540

common source was inheritance. Consider measureable characteristics of pairs of brothers (or sisters).

$$X = g + x,$$
$$Y = g + y,$$

where X and Y represent the variation of each brother from the mean of the species, g represents the variation of the parent from the mean of the species, and x and y represent variation of the brothers from the parent. Galton showed that the correlation between the pairs, X and Y, was the product of the correlation between the first member and g (X versus g) and the correlation between the second member and g (Y versus g),

$$r_{XY} = r_{Xg} * r_{Yg}.$$

The correlations $r(Xg)$ and $r(Yg)$ are the factor loadings of X and Y on the common, or inherited characteristic, g. The relation says if the correlation between parent and offspring is 0.50, then the correlation between brothers or sisters will be 0.25. In this case, $r(Xg)$ is the same as $r(Yg)$.

The second problem that interested Galton was also of interest to Scotland Yard—the identification of criminals. The existing method was based on twelve body measurements suggested by Alphonse Bertillon of France. The measures included height, length and breadth of head, length of forearm, length of middle finger, and the like. Bertillon suggested organizing records into three classes for each measure—leading to 3 exp(12) possible classes. But Galton pointed out that the measures were intercorrelated and not independent. Thus, as Burt remarks, Bertillon would find himself "in respect to many of his compartments in the position of Old Mother Hubbard." (Who found her cupboard was bare.) [6]

Galton later became interested in identification through fingerprinting, did research on its use, and promoted its adoption. Burt reports how a committee in 1893 was impressed by the merits of Galton's fingerprinting system in comparison to body measurements, adding that "the representatives of the police force were, I suspect, strongly influenced by the fact that a criminal would be very apt to leave his fingerprints on the scene of the crime; he would not leave his body measurements there." [7]

Around 1901 Spearman became interested in Galton's work, and how the concepts introduced there might be used in studying the nature of intelligence. Spearman had gathered data on tests of 36 schoolboys in England on various abilities, and had calculated the correlations among the tests. The results are in Table 17-6. The factor loadings for the principal component of this table (or matrix) of correlation coefficients is

Variable	1	2	3	4	5	6
Factor loading	0.9365	0.8940	0.8420	0.8042	0.7426	0.7207

Table 17-6. Spearman's Correlation Matrix (1904) for Six Tests.

Test		1	2	3	4	5	6
1	Classics		0.83	0.78	0.70	0.66	0.63
2	French	0.83		0.67	0.67	0.65	0.57
3	English	0.78	0.67		0.64	0.54	0.51
4	Mathematics	0.70	0.67	0.64		0.45	0.51
5	Discrimination of tones	0.66	0.65	0.54	0.45		0.40
6	Musical talent	0.63	0.57	0.51	0.51	0.40	

Now apply Galton's expression for the correlation of two variables based on their correlations with a common factor. We obtain, for example

$$r_{12} = r_{1c} * r_{2c}$$
$$= 0.9365 * 0.8940 = 0.8372.$$

The actual correlation of variables 1 and 2 (Classics and French test scores) is 0.83. Application of this relation for all the pairs of tests produces a table of correlation coefficients that would exist if the variability of each test were comprised of variability arising from two sources—the common factor variability and the variability peculiar to the specific test. The result is shown in Table 17-7.

While Spearman did not use the principal components method, the factor loadings were similar. The close agreement between the *matrix of rank one* and the original correlation table suggested to Spearman that one and the same systematic factor was present in all the tests—namely *general intelligence*. He wrote that "all examination, therefore, in the different sensory, school, or other specific intellectual faculties, may be regarded as so many independently obtained estimates of the one great common Intellective Function." [8]

Spearman became an advocate of the single-factor theory of intelligence. This theory was soon challenged in Britain by Godfrey Thomson. Thomson demonstrated that the appearance of single factors in data like Spearman's could be brought about also by selections of tests in which group factors (or abilities) existed among some of the tests, but no overall common factor

Table 17-7. Theory Matrix of Rank One from Spearman's Data.

Test		1	2	3	4	5	6
1	Classics		0.84	0.79	0.75	0.70	0.68
2	French	0.84		0.75	0.72	0.66	0.64
3	English	0.79	0.75		0.68	0.63	0.61
4	Mathematics	0.75	0.72	0.68		0.60	0.58
5	Discrimination of tones	0.70	0.66	0.63	0.60		0.54
6	Musical talent	0.68	0.64	0.61	0.58	0.54	

existed. Thomson did this in 1914 by simulating a system of group factors plus specific factors through multiple dice throws where some of the dice gave results in common for several tests and other dice represented spe ^ic factors. [9]

In the United States in the 1930s L. L. Thurstone (1887–1955) developed multiple factor analysis based on a technique called the *centroid method*, and Hotelling's contribution based on principal components appeared.

Criticism

Standardized tests for measuring IQ (Intelligence Quotient) in schoolchildren appeared very soon after Spearman's initial work. Among the early builders of IQ tests were James Cattell in the United States and Alfred Binet in France. There are critics who maintain that the principal effect of the early studies of inherited characteristics followed by the spread of IQ testing for student placement in Britain in the interwar years was to preserve a class-based, élitist structure in the British educational system. [10]

A number of interrelated points are involved in such criticism, which has its parallel in the United States. The overriding problem is the separability of the effects of heredity and environment. That it is not as simple as Galton's common factor example would suggest is brought out by an example given by J. B. S. Haldane.

> Let A be Jersey cattle and B Highland cattle. Let X be a Wiltshire dairy meadow and Y a Scottish moor. On the English pasture the Jersey cow will give a great deal more milk than the Highland cow. But on the Scottish moor the order will probably be reversed. The Highland cow will give less milk than in England. But the Jersey cow will probability give still less. In fact, it is very likely that she will give none at all. She will lie down and die. You cannot say that the Jersey is a better milk-yielder. You can only say that she is a better milk-yielder in a favourable environment, and that the response of her milk yield to changes of environment is larger than that of the Highland cow. [11]

It is claimed that group differences in IQ reflect cultural differences between groups and the testmakers, reinforced by processes of test and question selection which are validated by achievements as measured in the dominant social groups. In 1924, Godfrey Thomson discussed competing explanations for urban–rural differences that were of concern at the time.

> As we go out from the city into the rural districts, we find the average IQ of the children, as measured by the usual tests, becoming lower. Is this due to a real decrease in intelligence, or is it due to the fact that the country children, though intelligent enough, are less ready with verbal responses and are less familiar with the type of material used in the intelligence tests? And if the decrease in intelligence is real, is it caused by the lack of book learning, lack of vocabulary, lack of stimuli to the sharpening of wits which the town child finds; or is it caused by the pull of towns taking away the more imaginative and the mentally more alert? [12]

Ways of Life—An Example

In *Varieties of Human Value* Charles Morris presented a study of valuation of 13 *ways of life* by university students in the United States and five other countries. [13] The Ways are paragraph-length statements designed to reflect the cultural components of various world religions, ethical, and philosophical systems. Single-phrase summaries of the Ways (as given by Morris) are:

1. Preserve the best that man has attained.
2. Cultivate independence of persons and things.
3. Show sympathetic concern for others.
4. Experience festivity and solitude in alternation.
5. Act and enjoy life through group participation.
6. Constantly master changing conditions.
7. Integrate action, enjoyment, and contemplation.
8. Live in wholesome, carefree enjoyment.
9. Wait in quiet receptivity.
10. Control the self stoically.
11. Meditate on the inner life.
12. Chance adventuresome deeds.
13. Obey the cosmic purposes.

In Morris' basic studies, respondents were simply asked to indicate their liking of the various Ways by assigning a number from 1 (dislike it very much) to 7 (like it very much).

Factor Rotation

Figure 17-5 shows factor loadings for the first two principal components for preference scores of sophomore business school students for the Ways. These two components accounted for 4.33 of the total standardized variance of 13.00 for 13 Ways. One reason that more variance is not accounted for in the principal components is that the 13 Ways were selected to be statements of different underlying life styles.

In the graph we see a rotation of the factor loadings axes so that insofar as possible particular Ways load high on one rotated axis and not on the other. The result is a more clear and interpretable *factor structure*. Way 4 and Way 1 now characterize the negative and positive extremes of rotated Factor 1, while Ways 5 and 2 come close to representing the negative and positive extremes of rotated Factor 2. Since Way 5 is *enjoy through group* and Way 2 is *cultivate independence*, we might term rotated Factor 2 the group-independence dimension. It is not as easy to characterize rotated Factor 1.

There are various rotation schemes. The one used here is called *varimax* rotation because a rotation which maximizes the variance of factor loadings of

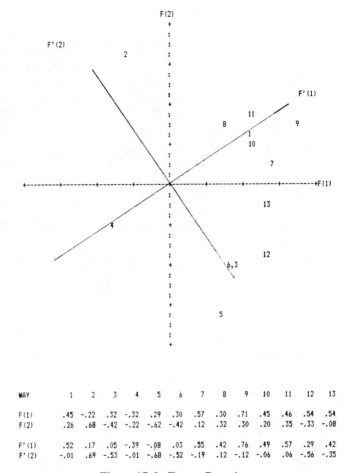

WAY	1	2	3	4	5	6	7	8	9	10	11	12	13
F(1)	.45	-.22	.32	-.32	.29	.30	.57	.30	.71	.45	.46	.54	.54
F(2)	.26	.68	-.42	-.22	-.62	-.42	.12	.32	.30	.20	.35	-.33	-.08
F'(1)	.52	.17	.05	-.39	-.08	.03	.55	.42	.76	.49	.57	.29	.42
F'(2)	-.01	.69	-.53	-.01	-.68	-.52	-.19	.12	-.12	-.06	.06	-.56	-.35

Figure 17-5. Factor Rotation.

original scales on the factors is sought. The total variance accounted for by the rotated factors is the same as that accounted for by the unrotated factors.

Semantic Analysis of the Ways

Charles Osgood and his associates in the 1950s developed a technique called the semantic differential in which they measured the meanings of a variety of concepts by pairs of adjectives such as good–bad, kind–cruel, soft–hard, and so on. They repeatedly found three factors of connotative meaning, which they labeled the *evaluative*, *potency*, and *activity* factors. [14] For example, *good–bad* is almost purely evaluative, *masculine–feminine* illustrates potency, and *fast–slow* illustrates the activity dimension.

In 1957, Osgood, Ware, and Morris obtained the ratings of 55 Stanford students on the Ways. In addition, they obtained the connotative meanings of the Ways by having the students score the Ways on 26 semantic differential (adjective) scales. When the Ways were used as variables and the mean semantic scale scores as observations, four factors appeared. These were labeled *dynamism, socialization, control,* and *venturousness.* [15]

In 1970, Morris and Small partly updated Morris' earlier study by having students then in college rate the Ways. They noted that students in 1970 scored lower on factors related to self-control and higher on factors reflecting withdrawal and self-sufficiency than did students in 1950. [16]

In 1973 the present author obtained semantic scores on the Ways for 31 University of New Mexico students. The mean semantic scores for the 13 Ways (variables) on 20 adjective scales (observations) were factored by principal components. Three factors appeared, which seemed to correspond well with the factors of dynamism, socialization, and control found earlier by Osgood, Ware, and Morris. The factors explained 87.4 percent of the sum of standardized variances of 13.0 for the Ways.

A Three-Dimensional Map

The factors form three measurement axes. When the scores of the adjectives on the three axes are standardized, the measurement space is essentially the inside of a globe with the origin at the point of intersection of the axis running from north to south pole and the plane of the equator. The factor scores for each adjective place that adjective in the three-dimensional space at a particular location having a specific orientation (or angle) with respect to each axis. Similarly, the factor loadings for each Way specify an inclination with respect to each axis. Both the factor scores (of adjectives) and factor loadings (of Ways) can be projected out to a unit distance from the origin onto the skin of the globe. The resulting picture of the adjectives and the Ways is much like the stars and the planets in a planetarium. They appear around us in their proper orientation, but distance from the origin is not emphasized.

Now we propose to rotate the measurement axis while the points stay fixed on the skin of the globe. We can rotate them anywhere we want as long as they stay at right angles to each other. We elected to run the polar axis through *true–false*. This adjective pair is almost pure *evaluative* dimension in Osgood's scheme. The equatorial plane is then perpendicular to the *true–false* axis. Adjectives on the equator have zero score on the *true–false* dimension. An example is *fast–slow*, which is nearly pure *activity* in Osgood's three-factor scheme.

To construct a two-dimensional picture of the globe with the adjective pairs

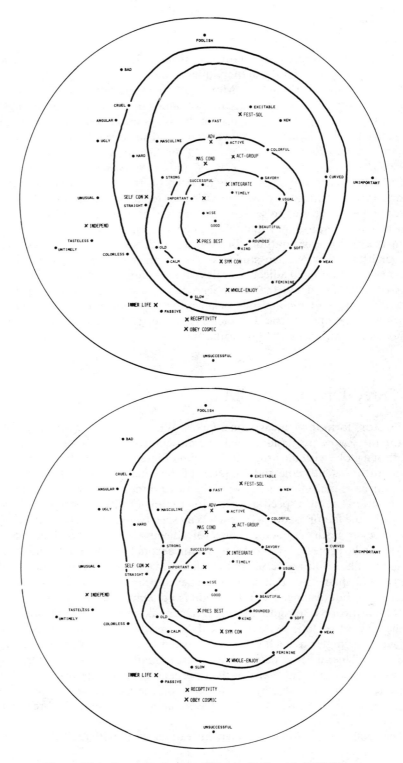

Figure 17-6. Semantic Space of Ways of Life with 1950 Values.

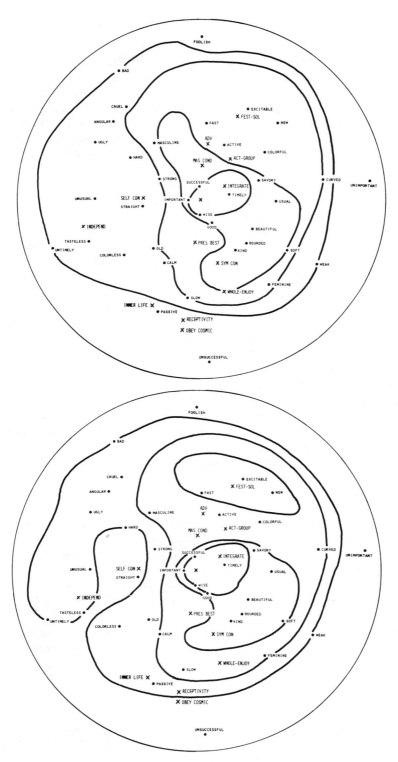

Figure 17-7. Semantic Space of Ways of Life with 1970 Values.

and 13 Ways of Life on the surface, we use a device of mapmakers—the equal azimuthal projection. The surface of the globe is seen as a circle whose center is at the north pole (true). The south pole (false) is represented by the circumference of the circle. The equator is a circle whose radius is half the distance from the north to the south pole. The projection preserves the true relation between distances that originate at or transverse the north pole, but other distances are distorted. Distortion increases as one goes south, and is extreme in the southern latitudes.

Figures 17-6 and 17-7 show the equal azimuthal projection of the semantic space of the Ways that has been described. Here we see descriptive properties like *important*, *successful*, and *wise* as close to *true*. *Fast-slow* forms an axis very nearly at right angles to *false-true*. Each pair of polar adjectives has one member in the northern hemisphere and its opposite in the southern. Each pair is half the diameter of the circle apart. All such distances transverse the pole (*true*) and are great circle distances covering half the circumference of the globe.

The Integrative Way (7) is seen as closest to *true*, while Way 12 (chance adventuresome deeds) is seen as *active* and Way 10 (control the self) is seen as *straight*. Way 3 (show sympathetic concern) is seen as *kind*.

Interpolated equi-value lines for the ratings of the Ways by male and female college students in 1950 and 1970 as reported by Morris and Small are also shown. These form a topographic map of valuation of the Ways, but also implicitly a valuation of the semantic properties seen in the Ways.

In general the highest valuations center on the Integrative Way. The valued properties are *true*, *wise*, and *timely*. The high value area shades off toward *good* and *beautiful* and toward *successful* and *important* in 1950, and the pattern of values for men and women are similar in 1950. In 1970 the valuations show some changes from 1950 and some contrasts between men and women.

First, the 1970 value topography is not as steep as in 1950. This can be described as reflecting more tolerant (or less discriminating) valuations of the Ways. Way 1 (preserve the best) and Way 12 (chance adventuresome deeds) are less valued. A result is that the second highest value area appears to tilt toward Way 3 (sympathetic concern). For men this next-highest value area extends also toward Way 6 (master changing conditions) in 1970. For women there is a separate plateau of next-highest value centering on Way 4 (alternate festivity and solitude). The implicit valuation of what is seen as *fast*, *excitable*, and *new* increased for women. Less value is attached to self-control by women than by men. It would seem that by 1970 women felt free to express values different from prevailing masculine values.

References

1. Francis Galton, *Memories of My Life*, Methuen, London, 1908, p. 302.
2. Cyril Burt, "Francis Galton and His Contributions to Psychology," *British Journal of Statistical Psychology*, **13** (May 1962), 31.

3. Harold Hotelling, "Analysis of a Complex of Statistical Variables into Principal Components," *Journal of Educational Psychology*, **24** (1933), 417–441 and 498–528.
4. Douglas F. Vincent, "The Origin and Development of Factor Analysis," *Applied Statistics*, **2** (1953), 116.
5. Godfrey H. Thomson, *The Factorial Analysis of Human Ability*, 5th edn., Houghton Mifflin, Boston, 1951, p. 94.
6. Cyril Burt, op. cit., p. 34.
7. Cyril Burt, op. cit., p. 38.
8. Charles Spearman, "General Intelligence," *American Journal of Psychology*, **15** (1904), 232. As reproduced in part in George A. Miller (ed), *Mathematics and Psychology*, Wiley, New York, 1964.
9. Herbert Solomon, "A Survey of Mathematical Models in Factor Analysis," in *Mathematical Thinking in the Measurement of Behavior*, (Herbert Solomon, ed.), The Free Press, Glencoe, IL, 1960, pp. 273–313.
10. Brian Evans and Bernard Waites, *IQ and Mental Testing*, Macmillan, New York, 1981.
11. Quoted in Evans and Waites, op. cit., p. 149.
12. Evans and Waites, op. cit., p. 169.
13. Charles Morris, *Varieties of Human Value*, University of Chicago Press, Chicago 1956.
14. C. E. Osgood, G. J. Suci, and P. H. Tannenbaum, *The Measurement of Meaning*, University of Illinois Press, Urbana, IL, 1957.
15. Charles E. Osgood, Edward E. Ware, and Charles Morris, "Analysis of the Connotative Meanings of a Variety of Human Values as Expressed by College Students," *Journal of Abnormal and Social Psychology*, **62**, No. 1 (1961), 62–73.
16. Charles Morris and Linwood Small, "Changes in Conceptions of the Good Life by American College Students from 1950 to 1970." *Journal of Personality and Social Psychology*, **20**, No. 2. (1971), 254–260.

CHAPTER 18

Jerzy Neyman

Jerzy Neyman (1894–1981) was born in Bendery near the border between Russia and Rumania. His father was a lawyer. Soon after his father's death in 1906, Neyman's family moved to Kharkov, in the Ukraine. He studied mathematics at the University of Kharkov, where the mathematician S. N. Bernstein introduced him to Karl Pearson's *Grammar of Science*. Having twice been rejected for military service, he graduated in 1917 and continued at the University to prepare for an academic career. He was married in 1919, imprisoned twice during the Russian revolution, received his master's degree in 1920, and subsequently lectured at the University in Kharkov. He went to Poland in 1921 in an exchange of nationals agreed to by Poland and Russia.

He secured a position at the National Agricultural Institute in Bydgoszcz as a senior statistical assistant, but returned to Warsaw after a year to be nearer to the university. In 1923 he became a lecturer in mathematics and statistics at the Central College of Agriculture. In 1924 he obtained his doctorate at the University of Warsaw, and continued to lecture at the Central College of Agriculture as well as at the university.

In 1925 Neyman was awarded a Polish Government Fellowship for study at University College in London. As Neyman tells it,

> These two guys—Sierpinski and Bassalik—came one day and said to me, 'You know, you're doing all this statistical stuff and nobody in Poland knows if its good or bad. So we are going to send you to London for a year to study with Karl Pearson, because you say he's the greatest statistician in the world. And if he will publish something of yours in his journal, then we'll known you're O.K. Otherwise—don't come back. [1]

The year Neyman arrived in London was the year of publication of R. A. Fisher's *Statistical Methods for Research Workers* and E. S. Pearson was an assistant in Karl Pearson's statistical laboratory. Neyman had met W. S.

Gosset (Student) on his arrival at the university. Gosset later arranged Neyman's introduction to R. A. Fisher. Jerzy and Lola Neyman soon became friendly with E. S. Pearson, a man about the same age as Neyman who had been raised in a very protected environment and had not yet emerged from the shadow of his father, "K. P."

E. S. Pearson to Gosset to Neyman

The first two great waves in the development of statistics were represented by the earlier work of Karl Pearson and the newer work emphasizing smaller samples that had begun with Gosset and had been greatly extended by Fisher. Conclusions were reached by comparing a test statistic with a critical value under the null hypothesis. E. S. Pearson began to wonder if there was some criterion apart from intuition that would tell the statistician what test statistic to use. He wrote to Gosset asking just how one should interpret rejection under Student's t-test. In reply, Gosset said that the only valid reason for rejection, no matter how large the t-value, was that some other hypothesis could explain what had been observed *with a greater degree of probability*.

The notion of an alternative hypothesis was the new element, and Pearson recognized how far-reaching it was. But he was unsure of his own mathematical abilities and knew that Gosset felt limited in this regard. Neyman was new to statistics and without attachments to either "K. P's" statistics or R. A. Fisher's. E. S. Pearson turned to Jerzy Neyman as "a foreigner whom I liked," and thus began the third wave of modern statistics. [2]

Neyman was to leave for a year of study in Paris, so that communication about their joint efforts had to be carried on by letter. Neyman's letters were saved by Pearson, but Pearson's letters have been lost. It appears that Neyman was slow to see what Pearson was driving at, but the first joint paper appeared in *Biometrika* in 1928, by which time Neyman was back in Warsaw with the Central Agricultural College.

Neyman–Pearson Tests

A good example of the contrast between the traditional test of a null hypothesis and the emphasis of Neyman–Pearson on decision-making was provided by Hogben in his 1955 text. [3] He asks us to think of a culture of fruit flies containing normal females and females with a sex-linked lethal gene. Normally the probability of a female offspring is 1/2, but if a female is a carrier of the lethal gene, the probability of a female offspring is 2/3. We observe that a particular female has 144 offspring. We are about to count the number of female offspring when we realize that we ought to have some rules for deciding whether the mother is normal or a carrier.

The count of females in 144 offspring will be our evidence. Common sense says that evidence of around half females will tend to support the conclusion that the female is normal, and evidence of around two-thirds female offspring will tend to support the alternative conclusion. But this can be quantified further.

We are going to observe a particular result—a count of females out of 144. From the binomial distribution we could figure the probability of this outcome

1. Under H0: the hypothesis that the female is normal

$$P(r_{\text{observed}}) = C_r^{144}(1/2)^r(1/2)^{144-r},$$

and

2. Under H1: the hypothesis that the female is a carrier

$$P(r_{\text{observed}}) = C_r^{144}(2/3)^r(1/3)^{144-r}.$$

Each of these probabilities is the likelihood of the sample result *given* one of the alternative hypotheses. It is easy enough to take the ratio of these likelihoods. The ratio of the likelihood under H0 to the likelihood under H1 for any evidence of the number of females is

$$L(\text{H0}:\text{H1}) = \frac{P(r|\text{H0})}{P(r|\text{H1})}.$$

In Table 18-1 we show these likelihood ratios for different numbers of females that might be observed. We see that the likelihood ratios change in the way suggested by common sense. Outcomes near one-half females (72/144) are more likely under H0 than under H1, and outcomes near two-thirds females (96/144) are more likely under H1.

At this point let us consider possible decision rules and their risks of error. A decision rule can be charcterized by the maximum number of female offspring that will lead us to conclude that the female is normal. If the observed number exceeds this we conclude that the female is a carrier. The following errors are possible.

Type I error: H0 is true and r exceeds $r(\text{max})$, leading us to conclude that H1 is true.

Type II error: H1 is true and r fails to exceed $r(\text{max})$, leading us to conclude that H0 is true.

The probabilities of Type I and Type II error can be calculated by summing the terms of the appropriate binomial distribution. The probability of a Type I error is called *alpha* and the probability of a Type II error is called *beta*. Table 18-1 shows these probabilities for the possible decision rules in our problem. They were in fact calculated by the normal distribution approximation to the binomials, but we need not go into that. It can be seen that the choice of a decision rule when the sample size is fixed (as here) is a matter of trade-off between the risks of Type I and Type II error. This is quite the same as the trade-off between producer and consumer risk in quality-control acceptance sampling that we saw in Chapter 16. We might think that a maximum acceptance number for H0 of 84 is a good decision rule, since both types of error are quite low under that rule.

What Neyman–Pearson showed was that the likelihood ratio was the appropriate criterion for defining a best test in the sense that for any level of alpha, there is no other test that will lead to a smaller beta than the test constructed on the likelihood ratio criterion. In particular they showed that this optimal feature was true for Student's *t*-test.

At a broader level Neyman–Pearson emphasized that attention should be given to the *alternatives* to the null hypothesis. This attention leads the investigator to consider the *power* of a test and the importance of determining the sample size needed to produce a desired power. We want now to illustrate these ideas.

The Power of a Test

The power of a test is the probability that the test will reject the null hypothesis when the alternative hypothesis is true. In the fruit fly example the alternative hypothesis (H1) is that the mother is a carrier ($P = 2/3$). We read from Table 18-1 that use of 84 as a maximum acceptance number for H0 would result in a 0.0212 probability of accepting H0 when H1 was true. The probability of rejecting H0 when H1 is true would then be $1 - 0.0212 = 0.9788$. This is the power of the test.

When there are multiple alternatives to the null hypothesis the power of the test depends on which alternative we are talking about. Let us return to the process control example of Chapter 8. There we were concerned with controlling a process which, when operating properly, produced a normal distribution of container fills with a mean of 1010.5 ml and a standard deviation of 3.0 ml. The analysis summarized in Figure 8-5 showed that the *control limits* based on a sample of nine observations were such that the chance of detecting a shift in the long-run mean fill to 1008.5 (a 2.0 ml shortage per fill) on the basis of a single sample following the shift was only 0.1517. The decision rule (sample size and control limits) did not have much *power* to detect a 2.0 ml shortage per fill.

If it was important to detect a shortage of 2.0 ml per fill on the basis of a single sample, it is possible to find out how large that sample must be. Suppose we want to be 95 percent sure to detect such a shift. This means the power of the test when the population mean is 1008.5 ml must be 0.95. The limitation on the probability of a Type II error is 0.05. In Table 8-1 we see that the *z*-score that gives a tail area of 0.05 in the normal distribution is 1.65. The control limits will be at plus and minus 3.0 SE $(\bar{x})$ from the control mean of 1010.5 ml.

Table 18-1. Errors for Alternative Decision
Rules for Choosing Between Two Hypotheses.

Max for H0	L(H0:H1)	Alpha	Beta
72	4996.153		
73	2494.331	0.4013	
74	1245.296	0.3372	
75	621.7145	0.2810	
76	310.3912	0.2266	
77	154.9629	0.0668	
78	77.36529	0.1401	0.0010
79	38.62456	0.1057	0.0018
80	19.28337	0.0778	0.0031
81	9.627230	0.0571	0.0052
82	4.806398	0.0401	0.0084
83	2.399596	0.0274	0.0136
84	1.197999	0.0188	0.0212
85	0.5981016	0.0122	0.0314
86	0.2986024	0.0078	0.0475
87	0.1490774	0.0049	0.0668
88	0.0744269	0.0030	0.0918
89	0.0371577	0.0018	0.1271
90	0.185510	0.0010	0.1660
91	0.0092616		0.2119
92	0.0046239		0.2676
93	0.0023085		0.3300
94	0.0011525		0.3936
95	0.0005754		0.4681
96	0.0002873		
97	0.0001434		
98	0.0000716		
99	0.0000357		

In terms of hypothesis testing the situation is

H0: Population mean = 1010.5 (process is in control)

Type I error limitation = alpha = 0.0013 (See Table 8-1).

H1: Population mean < 1010.5 (process is under-filling)

Type II error limitation = beta = 0.05 when shortage is 2.0 ml.

The sample size is found by

$$n = [z_{alpha} + z_{beta}]^2 [\sigma/d]^2,$$

where d is the critical change that we are concerned to detect. In the example,

$$n = [3.00 + 1.65]^2 [3.0/2.0]^2 = 49.$$

A sample of 49 will be needed. The lower control limit will be at

$$1010.5 - 3.0(2.0/\sqrt{49}) = 1009.21.$$

The z-score for this control limit when the population mean is 1008.5 (a shortage of 2.0 ml per fill) is

$$z = \frac{1009.21 - 1008.5}{3.0/\sqrt{49}} = 1.657.$$

Thus beta, when the population mean is 1008.5, is limited to 0.05 as was required.

Suppose we were willing to relax alpha, the probability of declaring that there was a shortage in mean fill when the process was really still in control. If this error risk were set at 0.05 and the Type II risk still set at 0.05, then the sample size required would be

$$n = [1.65 + 1.65]^2 \left[\frac{3.0}{2.0}\right]^2 = 25,$$

and the control limit would be set at a sample mean of

$$LCL = 1010.5 - 1.65\left(\frac{3.0}{\sqrt{25}}\right) = 1009.51.$$

The power curves for the two rules are shown in Figure 18-1. They are

Rule 1: $n = 49$, alpha $= 0.0013$, beta at 2.0 ml
shortage $= 0.05$; LCL $= 1009.21$.
Rule 2: $n = 25$, alpha $= 0.05$, beta at 2.0 ml
shortage $= 0.05$; LCL $= 1009.51$.

Of course, power of the tests to detect shortages is less than 0.95 for shortages less than 2.0 ml and more than 0.95 for shortages greater than 2.0 ml. These possibilities underscore the existence of multiple alternatives to H0.

Confidence Intervals

The construction of an interval estimate for a population value based on the data of a sample and the interpretation of the estimating procedure in terms of a relative frequency view of probability is due to Neyman. Historically there are two other approaches to interval estimation. One is through Bayes' theorem, which we turn to in the next chapter. It is the oldest method, and represents what is called *inverse probability* by many writers. R. A. Fisher rejected this concept and built a theory of estimation based on what he called *fiducial probability*. Neyman's *confidence interval* approach eventually came to contrast sharply with previous approaches. Neyman wrote in 1934.

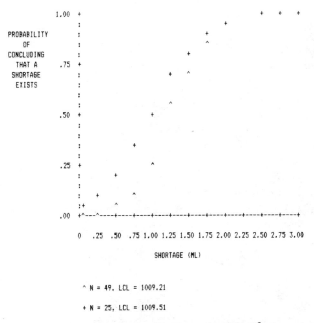

Figure 18-1. Power Curves for Detecting Fill Shortages.

The solution to the problem which I described as confidence intervals has been sought by the greatest minds since the work of Bayes 150 years ago. Any recent book on the theory of probability includes large sections concerning this problem. The present solution means, I think, not less than a revolution in *the theory* of statistics. [4]

We saw Bayes' theorem in Chapter 5 in Laplace's sixth *Principal of Probability*. We did not make a great deal of it there. It stated that

$$P(\text{Si}|\text{Ej}) = \frac{P(\text{Si and Ej})}{\sum[P(\text{S and Ej})]}.$$

Here, Si is a particular cause, or state, and Ej is an observed event. In the present context Si is a possible population value and Ej is an observed statistic. Bayes' theorem is a formula for finding the conditional probability of a population value *given an observed sample value*.

The numerator in Bayes' theorem is found using Laplace's fourth principle,

$$P(\text{Si and Ej}) = P(\text{Si}) * P(\text{Ej}|\text{Si}).$$

Here we see $P(\text{Ej}|\text{Si})$, the probability of the observed sample value *given a particular population value*. This is the kind of probability that we have been using. The sampling distribution of the (sample) mean (Chapter 8) is a whole collection of such probabilities with S fixed. The likelihood ratio illustrated earlier involved a fixed E and alternative Ss.

But $P(Ej|Si)$ must be multiplied by some pre-existing $P(Si)$ to obtain the probability of a state (or states) conditional on the observed sample value, $P(Si|Ej)$. This is where probability is *inverted*, or turned around. But where do these pre-existing, or *prior* probabilities come from? Fisher rejected the notion of prior probabilities and substituted his fiducial argument, which we will not go into. Neyman came to believe that Fisher was wrong—not in rejecting Bayes' theorem, but in his fiducial argument. To Neyman, the only legitimate probabilities were relative frequencies, and he sought an approach to interval estimation consistent with this view.

In Chapter 12 on Pearson, Gosset, and Fisher we saw that Gosset discovered the form of the distribution of ratios

$$t = \frac{\bar{X} - \mu}{s/\sqrt{n}}$$

in random samples from a normal population of X. The distribution was subsequently called Student's t-distribution.

In Figure 18-2 we show the results of taking repeated samples of size $n = 4$ from a normal population with a mean of 6.0 and a standard deviation of 1.0. We used the information in each sample to calculate

$$\bar{X} \pm 3.182s/\sqrt{n},$$

where 3.182 is the level of t for three degrees of freedom that leaves probability 0.025 in the upper right tail.

We see in Figure 18-2 that the 1st, 35th, and 36th ranges set up in this way fail to include the true population mean of 6.0. With a little thought we see why this is. If only 2.5 percent of repeated trials of t-values fall below -3.182 and only 2.5 percent fall above $+3.182$, we know only $2(2.5) = 5$ percent of repeated sample means in samples of size $n = 4$ will lie more than 3.182 t-multiples from the population mean.

An interval estimate constructed by $\bar{X} \pm t(0.025)s/\sqrt{n}$ is called the 95 percent confidence interval for the population mean because we know that the procedure by which the interval is constructed will lead to a range which includes the true mean in 95 percent of the cases when it is applied over the long-run. By the very nature of estimation, we do not know in which cases the estimate is incorrect. We only know the probability of correct and incorrect estimates for repeated trials of the method.

Statistical Issues and Personalities of the 1930s

The definitive paper by Neyman and E. S. Pearson entitled "On The Problem of Most Efficient Tests of Hypotheses" was read to the Royal Statistical Society by Karl Pearson on November 11, 1932. It had been favorably reviewed by R. A. Fisher. In June 1934 Neyman presented his paper entitled

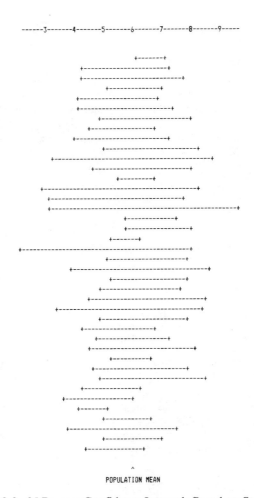

Figure 18-2. 95 Percent Confidence Intervals Based on Student's *t*.

"On Two Different Aspects of the Representative Method." The paper dealt mainly with important issues in survey sampling which we will review later, but it contained a short section on confidence intervals. Discussion of the paper centered on the confidence interval idea. In December, Fisher presented a paper, and Neyman was among the commentators. Neyman suggested that an alternative to Fisher's theory of inference might be possible, and Fisher responded by commenting that his and Neyman's methods seemed to give "a gratifying confirmation" because they gave identical results.

In February 1935 Neyman presented a joint paper on some aspects of experimental designs in agricultural field experiments. In discussing the paper, Fisher questioned Neyman's expertise in field experiments and suggested that he confused the questions of estimation and tests of hypotheses. The discussion was not pleasant, and E. S. Pearson questioned Fisher's infallibility in turn. This was the beginning of a continuing confrontation between Fisher and Neyman. Karl Pearson had yielded with difficulty to the new ideas of R. A. Fisher, and Fisher was not about to yield to Neyman's ideas.

Kruskal and Mosteller refer to Neyman's contributions to survey sampling as a "watershed." [5] In the nineteenth century, economic and social studies aimed for 100 percent coverage of relevant cases or intensive case studies of a selected subpopulation. These were mostly carried out by government statistical agencies. Beginning in 1895 a Norwegian, Anders Kiaer advocated the use of *the representative method*. This consisted of choosing districts, towns, parts of cities, etc., followed usually by systematic selection of ultimate units—houses, families, individuals, etc. For example, A. L. Bowley carried out a 1915 study entitled *Livelihood and Poverty* by securing data from a sample of one-in-twenty dwelling units in four selected British towns.

The ideal in the *representative method* is a miniature replica of the population that succeeds in duplicating several important characteristics such as the distribution of ages, marital states, and occupations of individuals. Neyman's 1934 article contains an account of a sampling failure by Corrado Gini and Luigi Galvani of the Italian Census. Gini and Galvani had to discard the records of an earlier Italian census so that room could be made for storing the current census. They decided to retain the data for 29 out of 214 administrative districts, and selected the districts so that they agreed with the population (census) values on seven selected characteristics. But they found that when other characteristics were examined, there were wide discrepancies between the sample and the population values. They decided that this was a condemnation of sampling in general, and advocated return to 100 percent coverage.

Neyman contended that the fault was not with sampling in general, but with *representative sampling* as then practised. He advocated explicit division of the population into *strata* in which the mean values of a control variable were the same, followed by a *random* selection of a pre-assigned number of districts from each strata. From this time, purposive representative sampling began to lose favor, although its use persisted in public opinion polling for another decade.

In the mid-1930s Frank Yates, Fisher's successor at Rothamstead, published a paper describing how experts performed in selecting samples by judgment. In an experiment involving wheat shoots and ears, experts tended to select ones that were too tall on the average and tended to avoid the extremely large and small ones. This general finding was confirmed in a similar study by Cochran and Watson, and in a later study by Yates involving selections from 1200 stones on a table. In the late 1930s Morris Hansen began his work at the United States Bureau of the Census on large-scale sample surveys. By this time Neyman had delivered an important set of lectures at the United States Department of Agriculture (1937).

We have already met W. S. Gosset (Student) as the brewmaster who found the form for the ratio of the error in the mean of a random sample from a normal population to the estimate of its own standard error (t), and whose comments set E. S. Pearson on a course which led to Neyman–Pearson hypothesis tests. He had introduced Neyman to Fisher by letter in 1927, commenting that "he is the only person except yourself that I have heard talk about likelihood as if he enjoyed it." [6]

Gosset had become interested in agricultural experiments through the concerns of the Guinness brewery with a program of the Irish Department of Agricutlure to improve the breeds and yields of barley. Through the 1920s he kept in touch with both Karl Pearson and R. A. Fisher through correspondence and frequent visits to London and Rothamstead. In addition to his work in Dublin with Guinness, he maintained a large correspondence with others concerned with practical statistical problems. He was a competent carpenter, fisherman, and sailor, and designed and built a number of boats for his own recreation. [7]

Among the questions that Gosset took to Karl Pearson in 1905 when he applied for a year of study at University College was one that involved correlation between paired observations. He had writen:

> Suppose observations A and B taken daily of two phenomena which are supposed to be connected. Let $A(1)$, $A(2)$, $A(3)$, etc. be the daily A observations and let $B(1)$, $B(2)$, $B(3)$ etc. be the daily B observations. . . . Then I form the two series $A(1) + B(1)$, $A(2) + B(2)$, etc., and $A(1) - B(1)$, $A(2) - B(2)$, etc., and find the P.E. (probable error) of each series. If they are markedly different, it is clear that the original series (sufficient observations being taken) are connected. . . . [8]

Gosset had not found a formula for measuring the connection, and Pearson provided it in the interview with him.

$$\mathrm{Var}(A + B) = \mathrm{Var}(A) + \mathrm{Var}(B) + 2r\sqrt{\mathrm{Var}(A) * \mathrm{Var}(B)},$$

$$\mathrm{Var}(A - B) = \mathrm{Var}(A) + \mathrm{Var}(B) - 2r\sqrt{\mathrm{Var}(A) * \mathrm{Var}(B)},$$

where r (underlined twice by Pearson) is Pearson's product-moment formula for the correlation coefficient,

$$r = \frac{\sum[(A - \bar{A})(B - \bar{B})]/n}{\sqrt{\mathrm{Var}(A) * \mathrm{Var}(B)}}.$$

Within two months Gosset was using correlation in reports for the brewery. By 1907, after his year at University College, he was using multiple correlation to assess the relative importance of different factors in influencing quality in brewing. [9] His second paper in 1908 (*Biometrika*) was entitled "Probable Error of a Correlation Coefficient." In it he used the same 750 samples of 4 that he used in the 1908 "On the Probable Error of a Mean" paper.

In his designs for agricultural experiments, Gosset put great reliance on difference testing that made use of the reduction in the standard error of a difference caused by positively correlated pairings. He would design field experiments in a *chessboard* system in which different treatments to be compared were placed in adjacent (rather than randomly selected) squares so that the correlation caused by similar soil conditions and drainage on pairs of yields to be compared were maximized. His advocacy of these *balanced* designs eventually drew him into a debate with Fisher, who advocated random assignment of plots to treatments in order to legitimize error formulas. It would seem that both concepts are important and are in fact combined in many good designs.

Gosset was able to maintain good relations with luminaries of his time who constantly fought each other. F. N. (Florence) David told Constance Reid

> I came to the Statistics Department (University College) as an instructor. Most of the time I was babysitting for Neyman, explaining to the students what the hell he was up to. . . . I saw the lot of them. Went flyfishing with Gosset. A nice man. Went to Fisher's seminars with Cochran and that gang. Endured K. P. Spent three years with Neyman. Then I was on Egon Pearson's faculty for years.
>
> They were all jealous of one another, afraid someone would get ahead. Gosset didn't have a jealous bone in his body. He asked the question. Egon Pearson to a certain extent phrased the question which Gosset had asked in statistical parlance. Neyman solved the problem mathematically. [10]

Gosset's letters reveal his modesty, candor, and humor. An example is a letter to R. A. Fisher about revising an earlier *t* table that had appeared in Karl Pearson's *Biometrika*. Gosset had wanted to offer the new table to Pearson in view of the earlier publication, but thought he should mention some small errors in the old table. Gosset tells Fisher that he mentioned the errors to "K. P." and left the old and the new tables with him. Gosset's letter continues:

> . . . when I came back on my way to Dublin I found that he agreed with me and that the *new* table was wrong. On further investigation both tables were found to be perfectly rotten. All 0.1 and 0.2 wrong in the fourth place, mostly it is true by 0.0001 only . . . . The fact is that I was even more ignorant when I made the first table then I am now . . . . Anyhow the old man is just about fed up with me as a computer and wouldn't even let me correct my own table. I don't blame him either.
>
> Whether he will have anything to do with our table I don't know . . . . It has been a rather miserable fortnight finding out what an ass I made of myself and from the point of view of the new table wholly wasted. However, I begin work again tomorrow. [11]

In fact, Gosset admired Karl Pearson, and E. S. Pearson feels that Gosset had his relationship of 25 years with "K. P." in mind when he wrote in 1933 that:

> the practical man, if he is not entirely foolish, talks over his problems with the professor, and the professor does not consider himself to be a competent critic unless he has some experience of applying the statistics to industry and has learned the difficulties of that application. [12]

E. S. Pearson wrote of Gosset that his "essential balance and tolerance" made him the ideal statistician. He was "a dependable person, quiet and unassuming, who worked not for personal reputation, but because he felt a job wanted doing and was therefore worth doing well." [13]

An issue highlighted by the depression of the 1930s was an alleged decline in the intelligence of the nation (United Kingdom, United States) that accompanied a tendency for birth rates of the affluent population to fall below those of the poor. The affluent were not even reproducing sufficiently to fill their own shoes, and gross statistics on crime and mental illness were increasing. The suggestion was that the geneological stock of the nation was deteriorating, and would continue to do so. Underlying these concerns were the two questions of the validity of intelligence measures and the effects of heredity and environment on so-called intelligence measures.

In 1930 an American psychologist, Carl Brigham, whose 1923 study had helped fuel the fears of national mental degeneration, reviewed the whole subject of intelligence testing. He came away skeptical of the entity called *general intelligence*, and closer to Walter Lippman's view that "intelligence is not an abstraction like length and weight; it is an exceedingly complicated notion which nobody had yet succeeded in defining." [14]

In England, two biologists who were also first-rate statisticians, Lancelot Hogben and J. B. S. Haldane, were critical of the views about deteriorating national stock. Hogben's group at the London School of Economics had mounted a serious research program to disentangle the relative weight of heredity versus environment (nature versus nurture) on intelligence measures. Hogben felt that the "increase" in mental deficiency reflected changes in criteria of classification and completeness of records. He noted that persons were not certified as feeble-minded unless they appeared at a police court, applied for poor-law relief, or were sent to special institutions for the retarded. There was no way to estimate mental deficiency "among the prosperous classess, where mental deficiency fades into the diplomatic service." Haldane said that observation of the bankruptcy courts showed that "a considerable number of the nobility are incapable of managing their own affairs." But, "they are not, however, segregated as imbeciles on that ground." [15]

Neyman in America

In December of 1933 E. S. Pearson invited Neyman to come to University College for three months in the spring. Earlier in the year Fisher had succeeded Karl Pearson as Galton Professor and Director of the Galton Laboratory, and E. S. Pearson had succeeded "K.P." as head of the Department of Applied Statistics. The two groups occupied adjacent floors but were hardly on speaking terms. Tea was served at 4 p.m. in the Common Room for Pearson's group, and then at 4:30 p.m. for Fisher's group when Pearson's troops were safely gone. Karl Pearson had separate offices for *Biometrika*, and apparently never entered his old building after his retirement as professor.

In the fall of 1934 Pearson secured a regular appointment for Neyman, which he held for four academic years. In the spring of 1937 Neyman made a trip to the United States. S. S. Wilks of Princeton and W. Edwards Deming of the Department of Agriculture were instrumental in arranging the trip. Deming arranged for a series of lectures and conferences at the Department of Agriculture. Deming was impressed with Neyman, and wrote to Raymond T. Birge, Chairman of Physics at the University of California at Berkeley, about him. Ultimately, and with the considerable involvement of Deming, an offer from Berkeley was accepted and Neyman arrived there in August of 1938.

Neyman spent the next few years establishing statistics courses and a laboratory at Berkeley. By 1942, however, the United States was at war. Younger talent was drawn away by the needs of the military and those who remained were involved as consultants and contractors to military and defense groups. Neyman's laboratory did work on bombing tables and later on high-level bombing of manuevering ships. The relative isolation of the west coast did not permit the same level of intensive involvement for Berkeley as Columbia and Princeton had.

In the fall of 1944 Neyman returned to England as a key person in a group organized by Warren Weaver to study Royal Air Force bombing data in order to prepare bombing strategies against Japan. He was back in the United States before Christmas. The war in Europe ended in June. Neyman began to organize a symposium, to be held at Berkeley in August, to contribute to the revival of academic work. The second day of the symposium was the day of the Japanese surrender aboard the U.S.S. *Missouri*.

Neyman retired in 1961 but continued to serve from year-to-year as a "professor recalled to active duty." He participated actively in the civil rights and anti-war movements during the 1960s. In January 1969 he was awarded the Medal of Science by President Johnson "for laying the foundations of modern statistics and devising tests and procedures which have become parts of the knowledge of every statistician." [16]

This recognition came at a time when Neyman's frequentist views were being questioned by a return again to the ideas of Bayes that had been rejected by both Fisher and Neyman. In reviewing the publication of a collection of Neyman's and Pearson's early papers, A. P. Dempster wrote,

> Neyman and Pearson rode roughshod over the elaborate but shaky logical structure of Fisher and started a movement which pushed the Bernoullian approach to a high-water mark from which, I believe, it is now returning to a more normal equilibrium with the Bayesian view. [17]

References

1. Constance Reid, *Neyman from Life*, Springer-Verlag, New York, 1982, p. 53.
2. Ibid., p. 62.

3. Lancelot Hogben, *Chance and Choice by Cardpack and Chessboard*, Parrish, London, 1955, p. 860.

4. Quoted in Hogben, op. cit., p. 898.

5. William Kruskal, and Frederick Mosteller, "Representative Sampling, IV: The History of the Concept in Statistics, 1895–1939," *International Statistical Review*, **48** (1980), 169–195.

6. Reid, op. cit., p. 59.

7. Launce McMullen, "Student as a Man," *Biometrika*, **30** (1939), 205–210.

8. E. S. Pearson, "Student as a Statistician," *Biometrika*, **30** (1939), 210–250.

9. Ibid., p. 216.

10. Reid, op. cit., p. 133.

11. Joan Fisher Box, "Gosset, Fisher, and the *t*-Distribution," *The American Statistician*, **35**, No, 2 (May 1981), 61–66.

12. Pearson, op. cit., p. 228.

13. Pearson, op. cit., p. 250.

14. Daniel J. Kevles, "Annals of Eugenics (Part III)," *The New Yorker*, October 22, 1984, p. 105.

15. Ibid., p. 107.

16. Reid, op. cit., p. 279.

17. Reid, op. cit., p. 274.

CHAPTER 19

The Bayesian Bandwagon

We ended our discussion of Laplace's principles of probability without exploring the tenth principle, which Laplace termed *moral hope*. We know this concept as *utility*.

Utility

Utility lies at an intersection of mathematics, psychology, and economics. Its beginnings for us precede Laplace's essay by some 60 years, and are found in Latin in *Papers of the Imperial Academy of Sciences in (St.) Petersburg*, Vol VI, 1738. by Daniel Bernoulli (1700–1782). Laplace credits Bernoulli for his tenth principle. Bernoulli's paper has been translated. [1]

Bernoulli considers whether a poor man who somehow obtains a lottery ticket that yields either nothing or 20,000 ducats with equal probability would be well advised to sell the ticket for 9,000 ducats, a sum below its expected value. He thinks so, and concludes

> ... the determination of the *value* of an item must not be based on its *price*, but rather on the *utility* it yields. The price of the item is dependent only on the thing itself and is equal for everyone; the utility, however, is dependent on the particular circumstances of the person making the estimate. Thus there is no doubt that a gain of a thousand ducats is more significant to a pauper than to a rich man though both gain the same amount. [2]

After further considerations Bernoulli states the generalization that became Laplace's tenth principle. It is that the utility resulting from any small increase in wealth *will be inversely proportional* to the quantity of goods already possessed. This means that the relation between money gain and utility will be of the shape shown in Figure 19-1 (also shown by Bernoulli). This shows the utility relation for someone we have come to call *risk-averse*. A gamble offered at fair money value will always be declined, because the expected utility gain will be less than the utility of the certain alternative. This is shown in the figure. Bernoulli, noting that both players in a fair game suffer an expected utility loss, comments that "indeed, this is nature's admonition to avoid the dice entirely." [3]

A consequence of the utility relation proposed by Bernoulli is that expected

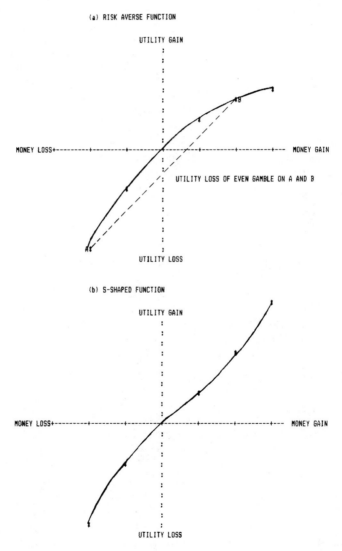

Figure 19-1. Utility Curves.

utilities are roots of products of constituent elements rather than arithmetic averages of the elements. For example, the 50–50 gamble for zero or 20,000 ducats to a man who already has 100,000 ducats has an expected *money* gain of

$$100(0.5) + 120(0.5) - 100 = 10 \quad \text{(thousand)},$$

but an expected *utility* gain of

$$[100^1 * 120^1]^{1/2} - 100 = 9.54 \quad \text{(thousand)}.$$

Bernoulli gives some further examples. One is whether a merchant who stands to gain 10,000 rubles by shipping goods from Amsterdam to St. Petersburg should insure them at a cost of 800 rubles when the probability is 0.05 of their loss at sea. On an expected money basis the insurance costs more than the risk avoided ($800 > 0.05 * 10,000$) and the insurance would not be bought. The expected utility without insurance is

$$[(X + 10000)^{19} * X^1]^{1/20}$$

and with insurance the utility of his fortune is

$$(X + 9200)^{20}.$$

Equating these two expressions yields a solution of 5043. If the merchant's current fortune is less than this he should insure the shipment; otherwise he should not.

A prospective insurer stands 19 chances in 20 of gaining 800 rubles and 1 chance in 20 of losing 9200 rubles. Solving

$$[(Y + 800)^{19} * (Y - 9200)^1] = Y^{20}$$

yields $Y = 14243$. A prospective insurer with a current fortune exceeding 14, 243 rubles should take on the insurance.

Bernoulli then brings his formulation of utility to bear on a problem which his cousin, Nicolas Bernoulli (1687–1759) had submitted to the mathematician, Pierre Montmort (1678–1719) in 1713. The problem is now known as the St Petersburg paradox.

Peter agrees to toss a coin until a head is obtained, and to pay Paul 1 ducat if heads occurs on the first toss, 2 ducats if heads first occurs on the second toss, 4, 8, 16, if heads first occurs on the third, fourth and fifth toss, and so on. What is the expected value of the agreement to Paul?

The answer, using expected money value, is

$$
\begin{aligned}
1d\,(1/2) &+ 2d\,(1/4) + 4d\,(1/8) \ldots + 2^{n-1}d\,(1/2)^n \\
= \quad 1/2 &+ \quad 1/2 + \quad 1/2 \ldots + \quad 1/2 \\
= \text{infinity.}
\end{aligned}
$$

But Nicolas Bernoulli had observed that any "reasonable" man would sell his opportunity in this game for less then 20 ducats. Daniel Bernoulli finds that if Paul had no fortune to begin with his chance would be worth 2 ducats in utility terms and 6 ducats if his fortune were 1000 ducats. His cousin informed him that the mathematician Gabriel Cramér (1704–1752) had proposed a solution in a letter (to N. Bernoulli) in 1728. Cramér's solution was to suppose that after some large prospective amount is reached, further doubling of money amounts would not further increase prospective utility. Whatever the validity of Cramér's solution, it does not seem as elegant mathematically as Daniel Bernoulli's.

Utility After Bernoulli

Bernoulli's formulation of utility was adopted by Laplace, but had no impact on economics or psychology for over 100 years. In 1850 a psychologist, Gustav T. Fechner (1801–1877) adopted the same formulation for the measurement of sensation. Known since as Fechner's law, the theory is that the sensation, S, is a function of the logarithm of the stimulus, R.

$$S = k * \log R.$$

A consequence of this relation is that the increase in sensation, ds, that accompanies a unit *arithmetic* increase in R will be constantly diminishing as R increases. Translated to economic terms in which R is units of a commodity and S is the utility of those units, the relation says that there will be a constantly *diminishing marginal utility* for additional units of the good. Diminishing marginal utility was taken up and made a touchstone of neoclassical economics by William Stanley Jevons (1835–1882). It was the view adopted by Alfred Marshall (1842–1924). However, Marshall and his followers used numerical utility as an introspective concept, and did not suggest that it was measurable in any operational sense. [4]

Marshall assumed that utilities held by different persons could be added together. This assumption bothered Francis Edgeworth (1845–1926) and later Vilfredo Paréto (1848–1923). The idea of the indifference curve, which implies how much of one good will be given up to obtain one unit of another good, was born, and became the foundation of a reformulation of consumer demand theory in the 1930s.

From the 1880s to the 1930s economics and psychology pursued separate paths, as seen in an account by Louis Thurstone. (1887–1955). Thurstone in the period 1927–1931 had made monumental contributions to attitude measurement, psychophysics, learning theory, and factor analysis. In 1931 he began to try to measure utility experimentally by asking subjects to compare the utilities of different bundles of commodities (for example eight hats and eight pairs of shoes versus nine hats and seven pairs of shoes). Thurstone wrote

> Around 1930 I had many discussions with my friend Henry Schultz, who was a mathematical economist. . . . In our discussions it became apparent that we were interested in different aspects of the same problem. Our discussions were clarified when we considered a three-dimensional model. The two coordinates for the base of the model were the amounts of the two commodities owned by the imaginary subject. The ordinates for the model were measures of utility. If

horizontal sections are taken in this model, we have a family of indifference curves, and when vertical sections are taken parallel to the base coordinates, we should have Fechner's law. [5]

A figure to visualize to appreciate Thurstone's model is the south-west quarter section of a beehive or inverted teacup.

In the 1950s some new ideas for the direct measurement of utility were put forward. John von Neumann (1903–1957) and Oscar Morgenstern (1902–1977), Princeton mathematicians, proposed what has come to be called the von Neumann–Morgenstern standard gamble technique. [6] In a standard gamble there are three goods, A, B, and C. It is known that A is preferred to B and B is preferred to C. If A and C be assigned utility values, then the utility of B is the expectation of a P chance of gaining A and a $1 - P$ chance of gaining C that will make the subject indifferent between the gamble and a sure prize of B. The method assumes that there is no utility attached to gambling per se. Suppose the utility of A is designated as 100 and the utility of C as zero. Then,

$$U(B) = 100P + 0\,(1 - P) = 100P,$$

where P is the probability that establishes indifference.

Milton Friedman and L. J. Savage (1917–1971) deduced from the fact that people buy both insurance and lottery tickets that some utility curves for money must have a reverse S-shape. [7] If one will buy insurance against large losses at more than the expected value of the loss (a necessity if insurance companies are to make any profit) and also pay more than the expected value for lottery tickets, the utility curve for money must swoop down rapidly at some point below the current wealth level and swoop up rapidly at some point above it. This is at variance with Bernoulli's and Fechner's idea of constantly diminishing marginal utility, and implies that people will take some gambles and refuse others, both for losses and gains. The S-shaped utility curve implies that gambles will be taken for modest losses and large gains in preference to the certain alternatives.

In 1951, Frederick Mosteller and Philip Nogee, reported on experiments in which individual utility curves were measured in ways implied by the standard gamble technique. [8] Later on researchers became interested in measuring utility curves for business and corporate decision-makers. [9], [10]

Bayes' Theorem

Thomas Bayes (1702–1761), the eldest son of a Presbyterian minister, followed his father into the ministry. By 1731 he had his own parish in Tunbridge Wells, near London. Not much is known of Bayes' early life. His education was private, and it is possible that he learned mathematics from De Moivre, who was known to have taught mathematics privately in London during that time. His mathematical attainments were recognized in 1742 by his election as a Fellow of the Royal Society. In response to a paper by Bishop Berkeley criticizing the methods of mathematics because mathematicians disagreed among themselves, Bayes wrote

> If the disputes of the professors of any science disparage the science itself, Logic and Metaphysics are much more to be disparaged than Mathematics; why, therefore, if I am half blind, must I take for my guide one that can't see at all? [11]

Bayes retired from the ministry in 1752 and died in Tunbridge Wells in 1761. In December 1763 Richard Price read a paper found among Bayes' effects to the Royal Society. The title of the paper was "An Essay Towards Solving a Problem in the Doctrine of Chances." In an introductory communication, Price recognized the originality of the contribution, calling the problem addressed "no less important than it is curious," and "a problem that has never before been solved." [12]

Bayes states the problem immediately in the essay

> *Given* the number of times in which an unknown event has happened and failed: *Required* the chance that the probability of its happening in a single trial lies somewhere between any two degrees of probability that can be named. [13]

He then reviews basic definitions and laws of probability. His Proposition 5, known to us as Bayes' theorem, is

> If there be two ... events, the probability of the second b/N and the probability of both together P/N, and it being first discovered that the second event has happened, from hence I guess that the first event has also happened, the probability that I am right is P/b. [14]

Here is an example. Suppose each day a coin is tossed. If the coin shows heads, a die is tossed. If the coin shows tails, a draw is made from a standard deck of playing cards. Now an *ace* in die tossing is the event that the die falls with one spot showing, while an *ace* in a draw from a standard card deck is the event that the ace of spades, the ace of hearts, the ace of diamonds, or the ace of clubs is drawn. Thus, if heads occurs on the coin, the probability of an *ace* is 1/6. But if tails occurs on the coin, the probability of an *ace* is $4/52 = 1/13$. The sequence of first and second events is shown in *tree diagram* form below.

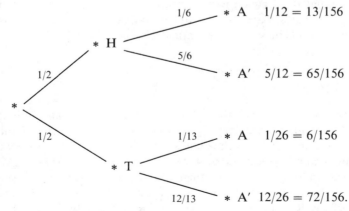

The probability of each sequence of events in the diagram is the product of the probabilities shown on the branches. These *joint* probabilities are shown at the right. The probability of the sequence head *and* ace is $(1/2)(1/6) = 13/156$, and the probability of the sequence tail *and* ace is $(1/2)(1/13) = 6/156$. The probability of an *ace* is the sum of the probabilities for the two *mutually exclusive* ways of getting an ace. This is

$$(13/156) + (6/156) = 19/156 = P(\text{Ace}).$$

In Bayes' terms, the first event in our example is the occurrence of a *head* and the second event is the occurrence of an *ace*. Bayes' proposition says that

$$P(\text{H}|\text{Ace}) = \frac{P(\text{Head } and \text{ Ace})}{P(\text{Ace})}$$

$$= \frac{P/N}{b/N} = \frac{P}{b}$$

$$= \frac{13/156}{19/156} = 13/19.$$

Think of these probabilities as expected frequencies for outcomes of the two events in a run of 156 days. On 13 of these days we *expect* heads *and* ace to occur, and on 6 of these days we *expect* tails *and* ace to occur. Therefore, the chances that heads occurred, *given that an ace occurred*, are $13/(13 + 6) = 13/19$ as above. Note that observation of an ace has changed our belief that heads occurred from a probability of 0.50 to $13/19 = 0.684$.

The Value of Information

We can use this same example to illustrate the basic ideas in decision-making and the value of information in a decision-making situation. Suppose that for the price of $1 each day we are able to receive $13 if heads and ace occurs and nothing otherwise. The joint probabilities in the tree diagram tell us that the expected return from our daily investment would be

$$(13/156)(\$13) + (143/156)(\$0) = \$(169/156)$$

or $1.0833. Since the chance at this expected return costs $1, our net return is $1.0833 − $1 = $0.0833 per day. The daily investment is a winning strategy compared to not investing.

Now consider that we are allowed to know whether an ace occurred before we have to decide whether to invest that day. We see from above that the probability of an ace is 19/156. If an ace occurs, we invest our dollar and take a 13/19 chance of gaining $13 (if heads preceded it) and a 6/19 chance of gaining nothing (if tails preceded). If an ace does not occur we do not invest because we gain nothing whether or not heads preceded it. This event occurs with probability 137/156. Our expected *net* return per day is now

$$(19/156)[-\$1 + (13/19)(\$13) + (6/19)(\$0)] + (137/156)(\$0) = \$150/156$$

$$= \$0.9615.$$

Knowing when an ace occurs allows us to avoid the mistake of investing $1 on those (137/156)ths of the days when there is no chance of a return. The information allows us to increase our net return per day by that expectation, or $1(137/156) = $0.8782. This amount is the expected gain from using the sample information to select the best strategy. Knowing if an ace has occurred lets us invest when an ace occurs and withhold investing when an ace does not occur. We would increase our expected net return by

$$\$0.9615 - \$0.0833 = \$0.8782.$$

This is called the *expected value of sample information.*

Break-Even Problems

A common class of problem is one where gain is a linear function of a mean or a proportion. The mean or proportion represents a state that we chose to regard as a random variable.

Our example is one from quality control. A process turns out lots of 1000 products each. The proportion defective in a large number of past lots has been distributed as follows:

Proportion defective	0.01	0.02	0.03	0.04	0.05	0.06	0.07	0.08	0.09	0.10
Proportion of lots	0.19	0.17	0.15	0.13	0.11	0.09	0.07	0.05	0.03	0.01

This distribution says that 19 percent of past lots have been 1 percent defective, 17 percent have been 2 percent defective, and so on. We have been selling lots for $6000, but a new buyer offers to pay $10,000 less $100 for each part found defective in later use. The table below shows the return under each payment schedule.

Proportion defective	Existing	New buyer
0.01	$6000	$9000
0.02	6000	8000
0.03	6000	7000
0.04	6000	6000
0.05	6000	5000
0.06	6000	4000
0.07	6000	3000
0.08	6000	2000
0.09	6000	1000
0.10	6000	0

There is a *break-even* point here. If we knew a lot was less than 4 percent defective we would sell to the new buyer, and if we knew a lot was more than 4 percent defective we would sell under the old basis. Right at 4 percent it would not matter. The mean, or expected proportion, from the probability distribution already presented, is 0.0385. Expected return under the two schedules is, then

Existing
$6000.

New
$10000 − 0.0385 ($100,000) = $6150.

It would be better to sell lots to the new buyer. But now it is proposed to test six parts from each lot. What is the best action to take given each possible sample outcome and what is the net value of following strategies based on this sample information?

In Table 19-1 we show the revision of the probability distribution of the proportion of defective parts for the possible sample occurrences of 0, 1, 2, 3, 4, 5, or 6 defects. Each row of Table 19-1(a) is the binomial distribution for $n = 6$ and the particular probability of a defective given under the heading STATE(S). These are then multiplied by the prior probabilities of states, $P(S)$, to obtain the joint probabilities in Table 19-1(b). Summing these joint probabilities for each sample outcome gives the probabilities for each sample number shown at the bottom of Table 19-1(b). Then, the revised probabilities of states are found by dividing the joint probabilities in each column by the total probability for the sample outcome. This is the application of Bayes' theorem, and the results are shown in Table 19-1(c).

At the bottom of Table 19-1(c) we show the mean of each of the revised distributions. If no defectives are found in the six parts, the revised expected proportion defective is 0.03519, and we would be better off selling to the new buyer. But if one or more defectives are found, the revised means are such that we could do better by selling to our old customers. Expected return from sampling and following these best strategies is

Sample outcome	Probability	Return	Probability × Return
0	0.79708	$6481	$5165.88
1	0.17714	6000	1062.84
2	0.02358	6000	141.48
3	0.00207	6000	12.42
4	0.00012	6000	0.72
5	*	6000	0.02
6	*	6000	0.00
		Expected return =	$6383.36

Table 19-1. Bayesian Analysis of Quality Control Problem.

(a) Probabilities of Sample Number Given States.

STATE(S)	$P(S)$	0	1	2	3	4	5	6	Total
0.01	0.19	0.94148	0.05706	0.00144	0.00002	0	0	0	1.0000
0.02	0.17	0.88584	0.10847	0.00553	0.00015	0	0	0	1.0000
0.03	0.15	0.83297	0.15457	0.01195	0.00049	0.00001	0	0	1.0000
0.04	0.13	0.78276	0.19569	0.02038	0.00113	0.00004	0	0	1.0000
0.05	0.11	0.73509	0.23213	0.03054	0.00214	0.00008	0	0	1.0000
0.06	0.09	0.68987	0.26421	0.04216	0.00359	0.00017	0	0	1.0000
0.07	0.07	0.64699	0.29219	0.05498	0.00552	0.00031	0.00001	0	1.0000
0.08	0.05	0.60636	0.31636	0.06877	0.00797	0.00052	0.00002	0	1.0000
0.09	0.03	0.56787	0.33698	0.08332	0.01099	0.00081	0.00003	0	1.0000
0.10	0.01	0.53144	0.35429	0.09841	0.01458	0.00121	0.00005	0	1.0000

(b) Joint Probabilities of States and Sample Outcomes.

STATE(S)	0	1	2	3	4	5	6	Total
0.01	0.17888	0.01084	0.00027	0	0	0	0	0.19000
0.02	0.15059	0.01844	0.00094	0.00003	0	0	0	0.17000
0.03	0.12495	0.02319	0.00179	0.00007	0	0	0	0.15000
0.04	0.10176	0.02544	0.00265	0.00015	0	0	0	0.13000
0.05	0.08086	0.02553	0.00336	0.00024	0.00001	0	0	0.11000
0.06	0.06209	0.02378	0.00379	0.00032	0.00002	0	0	0.09000
0.07	0.04529	0.02045	0.00385	0.00039	0.00002	0	0	0.07000
0.08	0.03032	0.01582	0.00344	0.00040	0.00003	0	0	0.05000
0.09	0.01704	0.01011	0.00250	0.00033	0.00002	0	0	0.03000
0.10	0.00531	0.00354	0.00098	0.00015	0.00001	0	0	0.01000
P(SAMPLE #)	0.79708	0.17714	0.02358	0.00207	0.00012	0	0	

(c) Revised Probabilities of States Given Each Sample Outcome.

STATE(S)	0	1	2	3	4	5	6
0.01	0.22442	0.06120	0.01161	0.00178	0.00024	0.00003	0
0.02	0.18893	0.10410	0.03989	0.01237	0.00338	0.00085	0.00020
0.03	0.15675	0.13089	0.07602	0.03572	0.01479	0.00564	0.00203
0.04	0.12766	0.14361	0.11237	0.07114	0.03969	0.02039	0.00989
0.05	0.10144	0.14415	0.14247	0.11393	0.08030	0.05211	0.03191
0.06	0.07789	0.13423	0.16090	0.15604	0.13338	0.10498	0.07797
0.07	0.05628	0.11546	0.16320	0.18664	0.18812	0.17460	0.15291
0.08	0.03804	0.08929	0.14581	0.19265	0.22433	0.24053	0.24337
0.09	0.02137	0.05707	0.10599	0.15927	0.21094	0.25724	0.29603
0.10	0.00667	0.02000	0.04173	0.07045	0.10483	0.14362	0.18568
Total	1.0000	1.0000	1.0000	1.0000	1.0000	1.0000	1.0000
Revised Mean	0.03519	0.04999	0.06127	0.06910	0.07470	0.07888	0.08212

The expected value of the sample information is

$$\text{EVSI} = \$6383.36 - \$6150 = \$233.86$$

because the information allows us to increase our expected gain by that amount.

We might ask what the limit on the possible value of sample information is. This question asks what *perfect* information would be worth in the decision situation that is faced. In our example perfect information would allow us to sell to the new buyer when we knew the proportion defective was 0.01, 0.02, or 0.03, and to sell to old buyers otherwise. We can find from the original distribution of lot quality and the schedule of returns under the two arrangements that the expected return under these circumstances is

0.19 ($9000) + 0.17 ($8000) + 0.15 ($7000) + 0.49 ($6000) = $7060.

Thus, the expected value of perfect information is

$$\text{EVPI} = \$7060 - \$6150 = \$910.$$

Bayesian Estimation—The Newsboy Problem

In both examples given so far, the decision problem was to chose between one of two possible states of nature. If the first state were chosen, a certain action was best and, if the second state were chosen, a different action was best. The role of sampling was to provide information on which the return from decision-making about states could be improved. In broad terms, chosing among states of nature is somewhat like classical hypothesis testing.

Recall that in classical Neyman–Pearson statistics, a second way of making an inference was estimation. This meant either a point estimate justified as best unbiased, maximum likelihood, or whatever, or a confidence interval usually calculated for large samples as a symmetrical interval about the sample mean or proportion.

A common problem in Bayesian estimation is typified by the "newsboy problem." A newsboy must buy papers daily from his supplier. Unsold papers may not be returned, and so are a complete loss. Daily demand is known but subject to random variation according to a probability distribution. The problem is to find the optimal number of papers to buy daily. This amounts to asking what is the optimal daily estimate of demand.

Suppose a particular out-of-town paper has demand at the newsstand conforming to a Poisson distribution with a mean of 8.0 papers. The newsboy buys the papers for 20 cents each and sells them for 30 cents. There are no expenses to charge against the margin of 10 cents per paper.

Consider the cost consequences of making errors in estimating demand. For every unit of underestimate, there will be a paper not stocked that could have been sold for a net gain of 10 cents. For every unit of overestimate, there will be a paper stocked but not sold, and the newsboy will have lost 20 cents spent to stock the paper. These consequences are sometimes called *opportunity losses*. They represent the cost, or loss from less than optimal decisions.

The optimal stock level can be found from Table 19-2. The table begins with the probability distribution of daily demand,

$$P(r) = e^{\mu}(\mu^r/r!),$$

Table 19-2. Optimal Demand Estimate in Newsboy Problem.

d	$P(D = d)$	$P(D < d)$	$P(D \geq d)$	Expected marginal gain
1	0.003	0.000	1.000	10.00 cents
2	0.011	0.003	0.997	9.91
3	0.029	0.032	0.968	9.04
4	0.057	0.089	0.911	7.33
5	0.092	0.181	0.819	4.57
6	0.122	0.303	0.697	$0.91 <$ Opt. Est. $= 6$
7	0.140	0.443	0.557	-3.29
8	0.140	0.583	0.417	
9	0.124	0.707	0.293	
10	0.099	0.805	0.195	
11	0.072	0.878	0.122	
12	0.048	0.926	0.074	
13	0.030	0.956	0.044	
14	0.017	0.973	0.027	
15	0.009	0.982	0.018	

where $e = 2.71828$ and $\mu = 8.0$ (see Chapter 7). Then, we find the expected opportunity loss for each unit increment in estimated demand (increase in the number of papers stocked). For example, let us first consider changing our estimate of demand from zero to one paper. The probability that demand is 1 or more is 1.0. It is certain that the first paper stocked will be sold, and the probability is zero that this paper will be unsold. The marginal expectation of gain from stocking this paper (increasing our estimate from zero to one) is

$$1.0 \ (10 \text{ cents}) - 0.0 \ (20 \text{ cents}) = 10.00 \text{ cents}.$$

We continue to think of increasing our estimate of demand by one unit. For the increase from 3 to 4, the expected gain is

$$0.911 \ (10) - 0.089 \ (20) = 7.33 \text{ cents}.$$

By the time we consider increasing our estimate from five to six papers the expected gain is only 0.91 cents, and the expected gain turns negative (a loss) when we consider increasing the estimate from six to seven papers. The expected return from stocking the sixth paper is only 0.91 cents, and the expected return from stocking the seventh paper is negative. The optimal estimate (or stock level) is six papers.

The newsboy problem is an example of a *linear loss* problem. The opportunity loss from underestimating is the loss per unit of underestimate times the numbers of units of underestimate, and the opportunity loss from overestimating is the loss per unit of overestimate times the number of units of overestimate. In the present example the loss per unit of underestimate was 10 cents (the potential gain from selling a paper) and the loss per unit of overestimate was 20 cents (the potential loss from failing to sell a paper). In general, the optimal estimate when opportunity losses are linear will be the highest value of the decision variable that has a probability of at least

$$Pc = \frac{Co}{Co + Cu}$$

of being equaled or exceeded, where Co is the increased loss per unit of overestimate and Cu is the increased loss per unit of underestimate. In our newsboy problem,

$$Pc = \frac{20}{20 + 10} = 0.67.$$

Reading down the column for $P(D \geq d)$, we find the highest d for which $P(D \geq d)$ exceeds Pc is 6, which is the optimal estimate. The common sense interpretation here is that when overestimating is twice as costly as underestimating, the best estimate is one for which the probability of an overestimate is half as great as the probability of an underestimate.

Subjective Probability

All of the examples so far in this chapter have involved initial probabilities that can be called *objective*. In our first example we took the probability for heads on the first event as $1/2$. In the second example the initial probabilities for proportion defective in a current lot came from long experience with the production process. But what if there is no apparent basis like an equally-likely assumption or experience data to go on? Many who adopt the Bayesian approach feel that in such cases the subjective belief of the decision-maker should be used for the initial probabilities.

A concept of probability as subjective belief really subsumes the ideas of probability based on equally likely cases and long-run relative frequencies. These become reasons for holding a degree of belief in certain cases. Advocates of a subjective or personalistic interpretation of probability say that the way to test a tentative belief (probability) is to ask if it corresponds to odds that you would give in a betting situation. [15] Suppose that you held a belief that the probability of heads on a particular coin was $2/3$. Then a payoff of \$1 should heads occur to \$2 should heads not occur should make you indifferent between taking either side of the bet, because

$$(1/3)\$2 = (2/3)\$1.$$

The test of belief is in actions that follow from it. If, on reconsidering, you preferred to bet on heads, then the probability (belief) you hold for heads must exceed $2/3$.

A necessary qualification for what has just been said is that your utility for additional money be proportional to the money gains over the range of bets (marginal utility of money is constant). If this were not so, we would not know whether your preference for the heads side of the bet meant that the utility of \$2 to you was less than twice the utility of \$1 as opposed to its meaning that the probability you held for heads was less than twice the probability you held for tails. In other words, you cannot use simple gambling choice behavior to measure both *utility* and *subjective probability* at the same time.

In the 1940s psychological studies of gambling [16] and horse-race betting behavior [17] had shown that people tend to overestimate the relative frequency of rare events and underestimate the relative frequency of common events. One explanation was that people's *subjective probabilities* did not agree with experience. Later Mosteller and Nogee pointed out that their results in utility measurement could be interpreted in terms of subjective probability. People may take long odds on lottery gains not because their utility curve swoops up, but because they really believe that their chances of winning exceed the actuarial odds. But how could one choose between these explanations?

This impasse had already been solved by Frank P. Ramsey (1903–1929) in 1926 and published in 1931. [18] Ramsey was a Fellow of King's College in Cambridge whose main interests were philosophy and mathematical logic. He had written a short book on *Foundations of Mathematics*, which J. M. Keynes said reflected "some of the best illumination which one of the brightest minds of our generation could give." [19] His early death left great promise unfulfilled.

Ramsey conceived the idea of an *ethically neutral* proposition with subjective probability 1/2. Consider the following pair of propositions:

(a) You receive $100 if the Dolphins win the Superbowl and $2 if the 49'ers win the Superbowl.
(b) You receive $100 if the 49'ers win the Superbowl and $2 if the Dolphins win the Superbowl.

If you have no preference between (a) and (b) then you must believe that the odds favoring the two teams are even, and you must not care (are ethically neutral) which team wins. Once an ethically neutral proposition with subjective probability 1/2 is found, it can be used as a standard to measure the utility of other propositions. [20]

Using Ramsey's idea, Davidson, Seigel, and Suppes carried out experiments designed to measure utility curves of Stanford University students. [21] As in the Mosteller–Nogee experiments, small monetary payoffs were used. These results further confirmed the existence of the S-shaped utility curves postulated by Friedman and Savage. [22]

Assessing Subjective Probabilities

We have seen that the relative frequency interpretation of probability in statistics prevailed through the years dominated by the ideas of Karl Pearson, R. A. Fisher, and Neyman–Pearson. Ramsey's contribution to subjective probability was virtually unknown in 1950. But there were other early dissenting voices. Among them were Harold Jeffreys (*Theory of Probability*, 1939) and Bruno de Finetti (*Foresight, its Logical Laws, its Subjective Sources*, 1937). Revival of interest in subjective probability and Bayesian decision-making and inference from 1950 to 1970 is attributed mainly to L. J. Savage (*The Foundations of Statistics*, 1954). Howard Raiffa and Robert Schlaifer of the Harvard Business School (*Applied Decision Theory*, 1961) were instrumental in spreading the Bayesian gospel to managerial decision-makers. [23]

Subjective probabilities are measures of personal belief. Consider the proposition *it will rain tomorrow*. If your probability is 1.0 you are certain of the proposition, if your probability is 0.5 you could not be more uncertain, and if your probability is 0.0 you are certain it will not rain tomorrow.

Since 1900 meteorologists have recognized the need to quantify the uncertainty involved in weather forecasts. The typical weather forecast is not mechanical, in that it is not based solely on an equation or set of rules specified in advance. The forecaster may look at a great deal of objective data (including forecasting equations), but winds up weighing all the evidence in a subjective fashion before making a prediction. A prediction may be in the form *it will rain tomorrow* or it may be in the form *the probability of rain tomorrow is 0.30*. The latter, known as a POP (probability of precipitation) forecast is a subjective probability. Since subjective probabilities are suggested for use by Bayesians when objective probabilities cannot be formulated, there is interest in methods for assessing subjective probabilities, and in how good subjectively assessed probabilities are.

Our first example is a forerunner of POP programs. In 1920 in response to requests by local farmers, a meteorologist named C. Hallenbeck developed a program for forecasting the chances of rain in the Pecos Valley in New Mexico using "percentages of probability." [24] The results of 123 forecasts are shown in Figure 19-2. We can use them to point out some features of subjective probability assessments.

First, it happened that the actual relative frequency of precipitation was 42.3 percent. The mean forecast percentage was about 45.2, so the overall forecasts seem reasonably close. But a result this good might be obtained using records on the percentage of days with rainfall for the same season. We must do something more than this. Goodness of the assessment is indicated by whether it rains on 25 percent of the days when the probability of rain is given as 0.25, half of the days when the probability of rain is given as 0.50, and so on. The graph emphasizes that the assessments look fairly good on this score as well, because the results tend to follow the ideal diagonal line.

In 1950 Brier suggested a scoring method for evaluating POP assessments. The score is the mean squared error of individual forecasts. Let $Y = 0$ stand for no rain, $Y = 1$ stand for rain, and X be the POP forecast given for the day. Then, the Brier score for N forecasts is

$$BS = \frac{\sum (X - Y)^2}{N}.$$

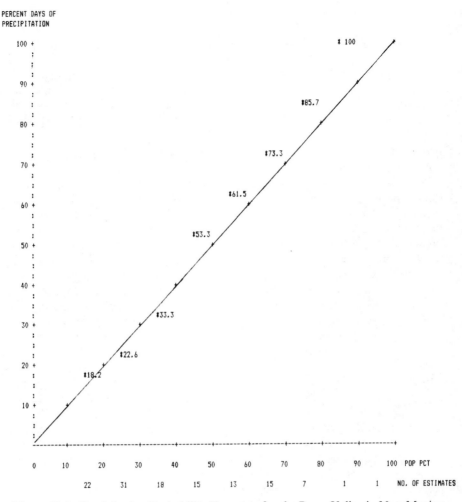

Figure 19-2. Precipitation Probability Forecasts for the Pecos Valley in New Mexico in 1920.

Later, the Brier score was partitioned into two terms. The first measures the correspondence between forecasts of a given POP and the actual relative frequency for those classes of POP. This is what we looked at when we asked whether it rained on 25 percent of the days when the forecaster gives the POP as 0.25. The second component of the Brier score measures the ability of the assessor to separate those situations for which the event will, from those for which it will not, occur. The first component reflects *calibration* of the assessor and the second reflects *resolution* of the assessor. In the Pecos example the resolution would be better if there were more forecasts at the lowest (0–9) and higher (80–100) POP percentages. The ideal probability assessor has high resolution and is well calibrated over the range of estimates. [25]

So much for a meteorological event that occurs or does not occur—like measurable precipitation. What about measured variables like maximum daily temperature or monthly rainfall? Here the assessor must give a *credible interval* statement like *the probability is* 0.50 *that the maximum temperature tomorrow will be between* 72 *degress and* 80 *degrees Farenheit*. This is a 50 percent credible interval. It is a central 50 percent credible interval if the 0.50 probability outside the interval is distributed 0.25 below 72 and 0.25 above 80 degrees. One can ask for other central credible intervals, such as the central 90 percent or central 99 percent intervals.

If less than 50 percent of actual occurrences were to fall inside 50 percent credible intervals established by an assessor, then the assessor is underestimating uncertainty. If more than half of actual occurrences fell inside the intervals, then the intervals are too wide, reflecting overestimation of uncertainty by the assessor(s) involved. Murphy and Winkler reported on experience in Denver and Milwaukee with credible interval forecasts of daily maximum and minimum temperatures. The percentage of 132 Denver temperatures outside the 50 percent credible intervals was 54.4 and the percentage of 432 Milwaukee temperatures outside the 50 percent credible intervals was 46.1. [26]

Results in areas other than meteorology are not as good. One must have situations in which the true values of quantities unknown to subjects are known at some point to the researcher so that quality of the assessment can be evaluated. A common experiment is to have subjects make credible interval statement is about *almanac* questions. Examples of almanac questions are

... What is the average annual rainfall in Jakarta?
... What is Reggie Jackson's lifetime major league batting average?

Lichtenstein reported on about ten almanac studies. The general tendency of assessors is to underestimate uncertainty, that is to be overconfident. The tendency ranged from mild to severe. [27]

Researchers have utilized a number of methods in obtaining subjective probability distributions for uncertain quantities. In the *fixed-interval* method the investigator quizzes the subject to obtain probabilities in fixed intervals, and in the *variable-interval* method the investigator tries to elicit an interval for a fixed probability. In a special case of this method the investigator tries to obtain the central 50 percent credible interval first. Then the upper and lower 25 percent areas are bisected to find the 12.5 and 87.5 percent points. Then these two extreme eighths can be bisected to find extreme sixteenths, and so on. Huber did a comprehensive review of these methods in 1974. His review agrees with Lichtenstein in the finding that subjective probability assessors do not attach sufficient variance to their probability distributions of uncertain quantities. [28] One is reminded of Yates' experiments of judgment samples of stones from a table mentioned in Chapter 15.

Bayesian Statistics

Once initial beliefs are in hand, whether subjective or not, the role of sample statistics for the Bayesian is to provide the likelihoods for use in Bayes theorem for revising one's belief on the basis of new evidence. Our quality control example showed how this works. When the initial distribution and the likelihoods take on particular forms, such as normal, there are formulas for determining the parameters of the revised distribution. These aspects of Bayesian statistics are covered in a number of texts. Some of the influential texts in this area are Schlaifer [29], Chernoff and Moses [30], Hays and Winkler [31], and Lindley [32].

References

1. Daniel Bernoulli, "Exposition of a New Theory on the Measurement of Risk," (translated by L. Sommer), *Econometrika*, **22** (1954), 23–36. Also reprinted in *Mathematics and Psychology* (George A. Miller, ed.), Wiley, New York, 1964.
2. Bernoulli, op. cit. (Miller), p. 37.
3. Bernoulli, op. cit. (Miller), p. 43.
4. William Fellner, *Probability and Profit*, Irwin, Homewood, IL, p. 85.
5. Louis Thurstone, *The Measurement of Value*, University of Chicago Press, Chicago, 1959, p. 17.
6. John Von Neumann and D. Morgenstern, *Theory of Games and Economic Behavior*, Princeton University Press, Princeton, NJ, 1947.
7. Milton Friedman and L. J. Savage, "The Utility Analysis of Choices Involving Risk," *Journal of Political Economy*, **56** (April 1948), 279–304.
8. Frederick Mosteller and Philip Nogee, "An Experimental Measurement of Utility," *Journal of Political Economy*, **59** (1951), 371–404.
9. C. Jackson Grayson, *Decisions Under Uncertainty*, Harvard Business School, Boston, MA, 1960.
10. P. E. Green, "Risk Attitudes and Chemical Investment Decisions," *Chemical Engineering Progress*, **59**, No. 1 (January 1963), 35–40.
11. G. A. Barnard, "Thomas Bayes—A Biographical Note," in *Studies in the History of Statistics and Probability*, (E. S. Pearson and M. G. Kendall, eds.), Griffin, London, 1970, pp. 131–133.
12. Thomas Bayes, *An Essay Towards Solving a Problem in the Doctrine of Chances*, in Pearson and Kendall, *op. cit.*, p. 135.
13. Ibid., p. 136.
14. Ibid., p. 139.
15. L. J. Savage, "Subjective Probability and Statistical Practice," in *The Foundations of Statistical Inference* (M. S. Bartlett, ed.) Methuen, London, 1962, pp. 9–35.
16. M. G. Preston and P. Barratra, "An Experimental Study of the Auction-Value of an Uncertain Outcome," *American Journal of Psychology*, **61** (1948), 183–193.
17. R. M. Griffith, "Odds Adjustments by American Horse-Race Bettors," *American Journal of Psychology*, **62** (1949), 290–294.
18. F. P. Ramsey, *The Foundations of Mathematics*, Harcourt-Brace, New York, 1931, pp. 172–184.
19. John Maynard Keynes, *Essays in Biography*, Macmillan, London, 1933, p. 296.

20. George A. Miller (ed.), *Mathematics and Psychology*, Wiley, New York, 1964, p. 66.
21. D. Davidson, S. Siegel, and P. Suppes, *Decision Making*, Stanford University Press, Stanford, CA, 1957.
22. Ernest W. Adams, "Survey of Bernoullian Utility Theory," in *Mathematical Thinking in the Measurement of Behavior* (Herbert Solomon, ed.), The Free Press, Glencoe, IL, 1959, p. 263.
23. John Aithison, *Choice Against Chance*, Addison-Wesley, Reading, MA, 1970, pp. 249–254.
24. Allan H. Murphy and Robert L. Winkler, "Probability Forecasting in Meteorology," *Journal of the American Statistical Association*, **79**, No. 387 (September 1984), 489–500.
25. Ibid., p. 491.
26. Allan H. Murphy and Robert L. Winkler, "the Reliability of Subjective probability Forecasts of Precipitation and Temperature," *Applied Statistics*, **26** (1977), 46.
27. Sarah Lichtenstein, *et al.*, "Calibration of Probabilities: The State of the Art," in *Decision Making and Change in Human Affairs* (H. Jungermann and G. de Zeeuw, eds.), Reidel, Dordrecht, Holland, 1977, pp. 275–324.
28. George P. Huber, "Methods for Quantifying Subjective Probabilities and Multi-Attribute Utilities," *Decision Sciences*, **5**, No. 3 (July 1974), 3–31.
29. Robert Schlaifer, *Probability and Statistics for Business Decisions: An Introduction to Managerial Economics Under Uncertainty*, McGraw-Hill, New York, 1959.
30. H. Chernoff and L. E. Moses, *Elementary Decision Theory*, Wiley, New York, 1959.
31. William L. Hays and Robert Winkler, *Statistics: Probability, Inference, and Decision*, Holt, Rinehart, and Winston, New York, 1970.
32. D. V. Lindley, *Introduction to Probability and Statistics from a Bayesian Viewpoint*, Cambridge University Press, Cambridge, UK, 1965.

CHAPTER 20

Observational Studies and Program Evaluation

The principles of experimental design were taught by R. A. Fisher and succeeding generations of statisticians and researchers. First among these principles is the random assignment of experimental material to treatments. This ensures that variables not controlled in the experiment do not introduce spurious effects and permits a measure of error separate from the effects of the treatments. This error is used as the basis for tests and estimates concerning treatment effects.

There are a host of situations that do not permit random assignment in socio-economic, educational, and health studies. Investigators must deal with the observations as they occur. We observe smokers and nonsmokers over a long period of time and look for differences in health and longevity that can be attributed to smoking. Some use the term *natural* as opposed to a *controlled* experiment. Perhaps it is better simply to call the uncontrolled case an *observational* study. [1]

Studies that are not true experiments in the sense of random assignment have been common in education since the 1920s, and much of the systematic thinking about analyzing the results is due to persons in this field. [2] Others have drawn their experience from public health and economic surveys. [3], [4] In the 1960s and 1970s requirements for evaluation of government programs in health, education, welfare, housing, and manpower led to many observational studies. The term *program evaluation* was applied to the endeavor to trace the effects of such programs. [5]

The Language of Cause

Philosophers of science have struggled with the concept of *causation*. We begin with an empirical approach to causation set forth in the Canons of Induction of John Stuart Mill (1806–1873). For Mill, event A can be said to cause event B if A precedes B in time and if B varies concommitantly with A. Simple examples are a magnet causing a nail to move or an electric switch causing a light to illuminate. But since other events, C, D, E, etc. may be occurring with and/or without A, it may not be easy to identify causes from observing events. Mill's Canons were an attempt to prescribe how to do this. For example, his Fourth Canon, called *the method of residues* stated that if one subtracts the effect of events previously shown to cause variations in an event of interest, the effect of remaining causes will be present as a "residue." We will see this principle at work in the interpretation of observational studies. [6]

C. West Churchman regards Mill as a *naive empiricist* who followed in the footsteps of John Locke (1632–1704). They believed that facts are primary and can be known, that laws (statements of causation) could be induced from observed facts, and that induction was complete in the sense that the laws permitted no exceptions. This position was modified by David Hume (1711–1776). Hume's position is identified as *statistical empiricism*, and holds that facts can only demonstrate laws (causal relations) in a probabilistic sense as tendencies based on experience. This was the position adopted by Pearson in *The Grammar of Science* when he emphasized classification of facts and recognition of fundamental sequences (cause–effect laws) as the basic task of science. [7]

In the search for causal relations it is useful to classify the variables of an observational study or survey into the following categories suggested by Kish. [8]

1. Explanatory variables. These are the variables involved in the cause–effect relationship that we wish to test.
2. Other *controlled* variables. These are variables (like C, D, E above) whose effects on the explanatory variables are controlled. Control can be achieved through
   (a) elimination of effects through the manner of selecting observations (by design).
   (b) elimination of effects by analytical means.
3. Other *uncontrolled* variables. Variables like C, D, E above whose effects on explanatory variables have not been eliminated or separately measured. The effects of these variables may be
   (a) so interrelated with the effects of explanatory variables that it impossible to make a clear attribution of cause and effect.
   (b) Unrelated to the effects of explanatory variables.

Kish emphasizes that in the ideal experiment there are no variables in Class 3(a). In observational studies there is more reliance on control through analytical means than through design because selection is not controlled. Again because natural happenstance can dominate over purposive design in an observational study, the possibility of confounding variables of Class 3(a) is a major concern.

If we are to establish the existence of a probabilistic law of cause and effect, the proposed relation must stand up to repeated tests that can conceivably find it false. Falsifiability has been emphasized in the modern philosophy of Karl Popper. By itself Mills' idea of concommitant variation is insufficient to establish a cause–effect relation, and no amount of testing can establish one with finality. The idea of falsifiability is that all conceivable alternatives to the explanation proposed must be eliminated before the proposed explanation is accepted. Cook and Campbell sum up the situation in these words.

> Thus the only process available for establishing a scientific theory is one of eliminating plausible rival hypotheses. Since these are never enumerable in advance, or at all, and since these are usually quite particular and require quite unique modes of elimination, this is inevitably a rather unsatisfactory and inconclusive procedure. But the logical analysis of our predicament as scientific knowers, from Hume to Popper, convinces us that this is the best we can do. [9]

Rival Explanations

Campbell and Stanley identified several classes of variables that could account for observed effects of explanatory variables in an uncontrolled set of observations. [10] The effects of these variables are frequently confounded with the explanatory variables. In reviewing them, it is useful to think of ways in which a simple *before–after* comparison on a criterion measure for an experimental group can be subject to alternative explanations. For example, consider an educational program whose objective is to foster more positive attitudes toward the world of work among seniors in a high school. Success is measured by change in scores on a paper-and-pencil attitude measurement instrument. The average change in scores from entrance to exit from the program, in addition to measuring the effects of the program, might be caused by

1. *History*. Perhaps an extraneous event, such as a widely seen television documentary or dramatic program, had more to do with any attitude changes than the educational program.
2. *Maturation*. If the educational program took place over the whole senior year, we have to consider the possibility that attitude change would have happened in any event as the seniors approached graduation.
3. *Testing*. It may happen that the test results on exit from the program are influenced by the fact that the students took the same or a similar test on entrance.
4. *Instrumentation*. If the attitude test contained free-form answers that had to be coded or scored, we have to reckon with the possibility that the judgments made in scoring or coding were not made in the same manner for the *after* as for the *before* test.
5. *Regression Effect*. If students were selected for the course because they had the *worst* attitude in a larger group tested, later scores would regress upward toward the overall mean attitude even if no real change in attitude

took place. How this happens is explained by the behavior of error factors in the measurements. Suppose two different, but equivalent (on the average) attitude tests are used, as might be the case in an attempt to avoid bias in the second test. Lack of perfect correlation of the tests reflects errors of measurement. Some of the low scores on the first test, used to select participants, will be low owing to these errors. Because the chance of the lowest scores having positive errors is less than the chance of those same scores having negative errors, the tendency will be for the average of those scores to rise on a retest even when no average change in attitude takes place.

Another important threat to validity can be seen if we change our proposed study. Instead of the before–after comparison of students in the course, we propose that the attitude scores of students on completion of the course be compared with the attitudes of students in one section of a required senior English course. The difference is to be attributed to the program. We now have a new threat to validity.

6. *Selection Effects.* In the proposed comparison there is an implicit assumption that the attitudes of the volunteers and English course students were the same to begin with. Because the groups are not selected at random, differences between them may be attributable to whatever the basis of selection was. In our case we might expect volunteers to have initially more positive attitudes than students in a representative section of a required course.

7. *Mortality.* When an intervention or its effects take place over time, there is the possibility for drop-outs and other disappearances of subjects. If this mortality takes place selectively, study validity can suffer. In the *after–only* comparison we would not be able to include the scores of persons who quit the course before completion. If these are students with poor attitudes, we already have a bias toward program "success" in comparing the scores for completers with the English section.

So far we have taken as our model of an observational study a simple *before–after* comparison of an experimental group and an *after–only* contrast between an experimental and a comparison group. A *before–after with randomly assigned control group* design would rule out these threats to validity. Subjects would be assigned randomly to experimental and control groups, and program effect would be measured as the *difference* between the change in scores for the experimental and the control groups. *History* and *maturation* effects would be eliminated because they would affect experimental and control group mean changes in the same way. The same could be said for simple *testing* and *instrumentation* effects. *Regression effects* would not occur with random assignment.

Let us now consider nonrandom assignment to an experimental and a comparison group. Again suppose the students in the special course were volunteers and that the study utilized for a comparison group a representative selection of senior students (taking care to exclude any volunteers). This pattern is a common one in many evaluation attempts.

The simple effects of history and maturation would be balanced out in this observational design as would be so in a true experimental versus control design. However, there are more subtle opportunities for mischief. Volunteers may already have a positive attitude toward work, while the comparison group is more average. An *interaction* between this selection bias and maturation could result in observing a negative "program effect." The comparison students might mature more rapidly than the already fairly mature volunteers, and the observed comparison group gain might well exceed the gain for those in the special course. Also *mortality* could affect experimental and comparison groups differently. If the course itself induces drop-outs of those least able to benefit, the change for the completing group is inflated in a way that is not balanced by comparison group change. Note that this could occur even with random initial assignment.

A similar kind of interaction between selection bias and regression effects plagues the nonequivalent comparison group situation. Again suppose that instead of volunteers, those with low attitude scores are somehow induced to participate in the course. A regression effect will cause their scores to rise on the average even in the absence of real program effects as explained earlier. If the comparison group is rather average to begin with, there will be no regression effect on their average score, and the comparison of gain for those in the special course with the control group could show a spurious "program effect."

This account does not exhaust the threats to validity that have been identified by Campbell and Stanley and Cook and Campbell. It does indicate the flavor and makes specific some of the kinds of rival hypotheses that must be capable of rejection before the assertion of a real program effect is justified. It is not sufficient to reject these rival hypotheses out-of-hand. The proper course is to anticipate them and provide controls through which their effects can be eliminated. We now turn to analytic methods for achieving control of these extraneous variables.

Analytical Control

The focus of analytic controls for removal of extraneous effects in observational studies is on the elimination of the effects of differences between the experimental and the comparison groups on the measure of program effect. Three approaches are

1. *Matching.* The idea here is to search a pool of potential comparison subjects to find one-on-one matches with members of the experimental group. The characteristics used for matching are those on which it is

suspected that overall differences between experimental and comparison groups exist. An alternative to one-on-one matching is to form subclasses by stratifying a population according to multiple characteristics. Then pairs are formed by random selection from within the strata.

2. *Analysis by Subclasses.* Both the experimental and comparison groups are classified into subgroups on the basis of factors related to program effects on which they might differ. Then, program effects are determined by experimental–comparison group differences *within* subgroups. In a technique called *standardization*, differences in the distribution of experimental and comparison groups by subgroups are eliminated by imposition of a standard distribution.

3. *Regression Methods.* The influence of extraneous factors on a measure of program effect can be eliminated through multiple regression. An equation that includes an indicator variable for experimental ($X = 1$) versus comparison ($X = 0$) group membership, along with measures of the extraneous variables is fit. The coefficient for group membership in the multiple regression equation measures the program effect *net* of the extraneous factors.

There are limitations to each of these methods. Individual matching is often practically difficult to carry out. Analysis by subgroups becomes unwieldy with three or more classification variables, and sampling errors increase as subgroup sizes diminish. Regression analysis can become untrustworthy in removing extraneous effects when the extraneous factor and causal variable being tested are too highly interrelated. We now turn to some examples to clarify the nature and limitations of the methods.

Auto Death Rates and Safety Inspections

Edward R. Tufte gives an example of automobile accident deaths per 100,000 population in each of the 49 states. [11] Some states had compulsory auto safety inspections and some did not. Tufte's data showed that the average 1966–1968 death rate for 31 states without inspection was 31.9, while the average rate in 18 states with inspections was 26.1. Should we attribute the difference to a beneficial effect of safety programs?

Tufte presents data that show that automobile accident death rates are strongly associated with differences in population density among the states. Western states such as Wyoming, Nevada, and New Mexico are thinly populated. High-speed accidents on lightly traveled highways produce high ratios of fatalities to injuries. In more densely populated states, auto accident death rates per 100,000 population are normally lower. If, on balance, the more densely populated states were more disposed to adopt compulsory inspections, their death rates might be lower even if inspections had no effect.

Subgrouping is one way to check the rival hypothesis that the death rate

Table 20-1. Deaths Rates from Auto Accidents per 100,000 Population.

Condition	States with low density		States with medium density		States with high density	
	Average	N	Average	N	Average	N
No inspection	38.5	9	31.5	16	23.6	6
Inspection	34.9	3	28.4	9	18.3	6

difference is attributable to differences in population densities. Table 20-1 presents comparisons for subgroups of states according to population density. Here we see that it is true that states with high and medium population density are more disposed to have inspections than states with low population density. More importantly, we can now compare the averages with and without inspection for groups of states with similar density. We see differences in favor of inspection of 3.6, 3.1, and 5.3 for low, medium, and high density groups, respectively. Note that these are all smaller than the overall difference of $31.9 - 26.1 = 5.8$ given earlier. Thus, some of the gross difference seems explainable by the density factor. Differences remain that favor inspection.

Here we can ask a question that leads to the idea of standard weights. The overall figure for states without inspection is a weighted average of 38.5, 31.5, and 23.6. The weights are the numbers of states. We can ask what the average for the no inspection group would have been with these same weights. We have

$$\frac{38.5(9) + 31.5(16) + 23.6(6)}{9 + 16 + 6} = 31.9$$

for states without inspection, and

$$\frac{34.9(9) + 28.4(16) + 18.3(6)}{9 + 16 + 6} = 28.3$$

for a similarly weighted average with inspection. The standard set of weights removes the effect of differing subgroup numbers between the experimental and comparison groups. But we find a difference of 3.6 remains that favors inspections.

Laura Langbein [12] presents an auto accident death rate analysis based on data presented by Colton and Buxbaum. [13] The criterion measure is age-adjusted motor vehicle accident deaths per 100,000 white males 15–64 years old. Death rates are found initially for each age group by states, i.e., 15–24, 25–34, 35–44, 45–54, and 55–64. Then the overall figure for each state is found by applying a standard set of weights (population distribution) to the age-specific rates for the state. This *standardization* procedure removes the influence of different age distributions on the accident death rates. Thus a rival hypothesis that state-to-state differences in auto accident death rates are attributable to different age distributions in the states is ruled out.

Age-adjusted death rates for 1960 are examined to find if a difference can be attributed to inspection programs. The method employed is multiple regression. Coefficients are found for the regression

$$Yc = a + b(X) + c(Z),$$

where b is the coefficient associated with states without ($X = 0$) versus states with ($X = 1$) inspection programs, and c is the coefficient associated with a unit difference in the logarithm of population density. The equation was found to be

$$Yc = 84.71 - 6.34(X) - 9.08(Z),$$

with the coefficients for both programs and density significant. This equation is, in a manner of speaking, two equations. For states without inspection ($X = 0$) we have

$$Yc = 84.71 - 9.08(Z)$$

and for states with inspection we have

$$Yc = 84.71 - 6.34(1) - 9.08(Z)$$
$$= 75.63 - 9.08(Z).$$

These equations are plotted in Figure 20-1. The regression *model* used says that the effect of density is the same for states with and without inspection programs (the regressions are parallel). The vertical distance between the lines

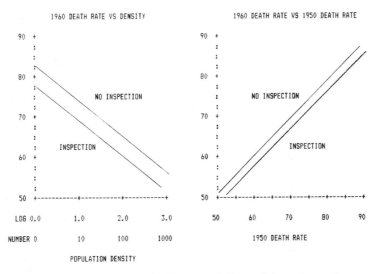

Figure 20-1. Regressions for Program Effects of Auto Inspections.

is the coefficient of X, and represents a constant difference attributable to inspection in states with comparable density.

It may be objected that inspection programs are adopted in states that have lower auto accident death rates, and that the difference favoring inspection programs above is the result of a favorable selection bias. To check on this possibility a regression was run using 1950 death rates as Z and employing X as the program variable. The results were

$$Yc = 4.43 - 2.32(X) + 0.93(Z).$$

The program difference of 2.32 was insignificant and the coefficient for 1950 death rates was highly significant. We thus have support for a rival hypothesis that destroys our earlier evidence favoring a program effect. This equation is shown in Figure 20-1.

A regression was run that includes both population density and 1950 mortality rate. Both of these extraneous variables were found significant, and the regression coefficient for program effect was insignificant.

A danger with regression adjustment is undercorrection for the effects of the extraneous variables. Virtually all the imperfections in the method tend to this result. Interactions between this error and selection bias can cause a compensatory social program to show a measured negative effect when the true correction would have shown positive effects. [14]

Systems of Relationships

In 1927, E. J. Working published an article entitled "What Do Statistical 'Demand Curves' Show?" [15] The major point of the paper is illustrated in Figure 20-2. Panel (a) shows a supply and a demand curve as often drawn by economists. The supply curve shows the quantities of a good that would be brought into a market by suppliers at different prices, and the demand curve shows the quantities that would be purchased by buyers at different prices. In a competitive market price tends toward the level that equates supply and demand at the intersection of the two curves.

Suppose we talk about yearly average prices and quantities supplied for an agricultural commodity. As shown in Panel (b), if the demand curve remains the same over time and the supply curve moves in response to changing weather, crop conditions and costs, then the variation in the intersections of supply and demand over the years will trace out the demand curve. Correspondingly, in Panel (c) we show a situation where demand is determined by outside (exogenous) factors and the supply curve does not change from year to year. The intersections trace out the supply curve. Where both demand and supply vary, the resulting curve is a historical relation of price versus quantity intersections, but it does not trace out the shape of either the demand or the

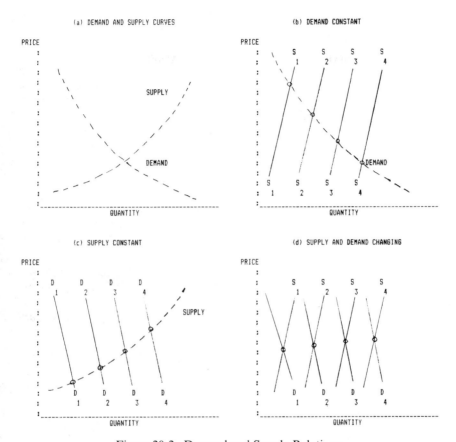

Figure 20-2. Demand and Supply Relations.

supply curve as the economist understands them. It shows neither *structural* relation.

In demand studies for agricultural products such as meat, poultry, eggs, feed grains, and certain fresh fruits and vegetables using ordinary least squares analysis (OLS), price is used as the dependent variable, and production along with *demand shifters* such as income are the X, or predictor variables. The demand shifter(s) are the variables which move the demand curve to the right or left. They are like the extraneous variables in an evaluation study. We want to adjust for their effects and obtain a single *structural* relation for the demand curve. The relation usually sought is the slope of the demand curve, often in terms of the percentage increase in quantity accompanying a 1 percent decrease in price.

If both the demand and the supply curves are shifting, their intersections will trace out neither *structural* relation. In 1943 Trygve Haavelmo introduced a simultaneous equation method of determining the responses of both demand and supply to changing price when neither the demand or the supply curve stays fixed. The method involves an initial demand equation in which quantity is a function of price and a demand shifter like income, and an initial supply equation in which quantity is a function of price and a supply shifter like costs. These equations are manipulated to form equations for price and quantity in terms of the two exogenous variables whose solution by least squares will in turn provide a solution to the original structural equations. [16] By the mid-1950s a variety of econometric methods for estimating parameters of systems of equations had come into use.

In 1921 a biologist, Sewell Wright, had published a paper entitled "Correlation and Causation." [17] A later paper in 1934 called "The Method of Path Coefficients" contains exposition and examples of what has since come to be known as *path analysis*. Path analysis deals with systems of cause–effect relations. Wright regarded the method as a way of evolving a consistent interpretation of a system of correlations. [18]

To appreciate the ideas in path analysis it is useful to restate equations for correlation among three variables in a particular form. If one of the variables is regarded as dependent and the other two as predictors of the first, the regression equation is usually written as

$$\text{est } Y = a + b_{1Y}X_1 + b_{2Y}X_2.$$

If the variables are all expressed in standard scores, the coefficients in the equation

$$\text{est } y = B_{1Y}x_1 + B_{2Y}x_2$$

are obtained by solving

$$B_{1Y} + B_{2Y} * r_{12} = r_{1Y},$$
$$B_{1Y} * r_{12} + B_{2Y} = r_{2Y}.$$

Here the B-coefficients are the standard deviation units change in Y per standard deviation change in X, and the rs are the correlation coefficients between the indicated variables. The Bs are sometimes called *beta*-coefficients.

These two equations can be represented by the path diagram in Figure 20-3. Here the Bs are directional because they indicate the effect on Y per unit (standard deviation) of X. The correlation coefficients are not directional because they can be read either way. That is, the correlation coefficient between the two predictor variables is *both* the standardized slope in $X(1)$ per standard deviation of $X(2)$ and the standardized slope in $X(2)$ per standard deviation of $X(1)$.

Notice that a rule can be deduced for writing the equations just presented. To write the equation for $r(1Y)$ for example, one adds the products of coefficients that appear on the path going forward from 1 to Y to the products of coefficients that appear on the path going backwards from 1 and thence to

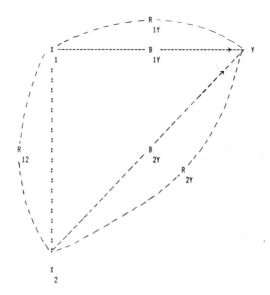

Figure 20-3. Basic Path Diagram and Coefficients.

Y. That is,

$$\text{Forwards:} \quad B_{1Y},$$
$$+ \text{Backwards:} \quad r_{12} * B_{2Y}.$$

Thus,

$$r_{1Y} = B_{1Y} + r_{12} * B_{2Y}.$$

The beta-coefficients have long been used as measures of the strength of effect of different predictor variables on a dependent variable. In path analysis the beta-coefficients are given the symbol p for *path coefficients*. They measure the strength of directional connections in the path diagram.

The path diagram itself embodies a view of causal connections suggested by theory. If we believed that $X(1)$ could affect Y only by inducing changes in $X(2)$, we would not show a directed path between $X(1)$ and Y. An example is where an individual's knowledge about employment opportunities, $X(1)$ was viewed as causally connected with the individual's wages (Y) only through the effect of knowledge on the person's motivation to get ahead. There would, of course, be other variables affecting motivation. In a case like this motivation would be said to be an *intervening* variable between knowledge and wage level. Wright emphasized that the method . . . "usually involves a working back and forth between tentative hypotheses based on external information and the set of correlations until as simple and reasonable a conclusion has been reached as the data permit." [19]

Discriminant Analysis

So far we have discussed the effects of alternative treatments or programs in terms of a measured criterion of success. Examples were highway accident death rates per 100,000 population and a measure of attitude towards the world of work. In other cases the measure of program success may be just that—the program worked or it did not. A manpower training director may be concerned only with whether each trainee secures a job or not on completion of the training. The measure of success of a college remedial academic skills program may be simply whether the student clients successfully remained in college, or ultimately whether each student graduated.

When the dependent variable in a study is a nominal, or categorical variable, and the predictor variables are measured variables, regression analysis is inappropriate. The reason is that a normal distribution of residual variation from the regression line required for proper use of standard errors of the regression coefficients is not present. The appropriate method is *discriminant analysis*.

A classic example of discriminant analysis was given by R. A. Fisher concerning different varieties of iris. Fisher's example dealt with classifying three varieties of iris (setosa, versicolor, and virginica) on the basis of four measurements (sepal length, sepal width, petal length, and petal width). [20] We will consider distinguishing between iris setosa and iris versicolor on the basis of measurements of sepal length and sepal width. The data for 16 iris setosa and 17 iris versicolor are shown in Figure 20-4.

In Figure 20-4 we see that both the sepal lengths and sepal widths overlap for the two varieties of iris. A vertical line is drawn in at the mean sepal width. If this line were used to distinguish setosa from versicolor, we can see that two setosa would be misclassified as versicolor and four versicolor misclassed as setosa. If mean sepal length were used (horizontal line), the situation is just reversed, with classification errors made for two versicolors and four setosas. One can move these lines about and classification errors still result. On the other hand, the diagonal line, labeled *cutting plane*, which is determined by discriminant analysis, does separate the varieties correctly, with the setosas lying above and the versicolors below the line. The line at right angles to the cutting plane is the axis for a discriminant function, which is a linear combination of the two original measures, sepal width and sepal length. The discriminant function here is

$$D = 1.387 \text{ (sepal length)} - 2.417 \text{ (sepal width)}.$$

This is the linear function which maximizes the standard score separation between the two varieties.

A Discriminant Analysis of Selection Factors

Earlier the point was made that participants in a program may differ from nonparticipants in many ways. Therefore a comparison of outcomes for

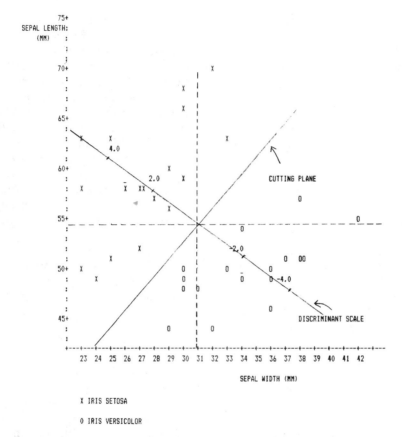

Figure 20-4. Discriminant Analysis of Iris Data.

participants with nonparticipants must take these differences into account. Use of cross-classification and regression analysis were two of the ways of taking account of these extraneous variables. A discriminant analysis of participants versus nonparticipants in a program can help to find these variables.

During the 1970s nutritional centers which served meals and dispensed nutritional information for the elderly were operated under contracts from the Department of Health, Education, and Welfare. As part of a study to develop the design for an ongoing national evaluation of these centers a pilot study was made in three areas in which centers were operating. In this study program participants in the three areas were compared to a random sample of qualifying nonparticipants from the same areas. [21]

Initially a set of 44 variables that might be related to whether persons elected to take meals at the centers was drawn up. The variables described demographic characteristics, income, food purchasing and preparation, and physical and psychological health. After a preliminary analysis the list of variables was reduced to 19, which were included in a stepwise discriminant analysis of 73 participants and 68 nonparticipants for whom complete information was available. A stepwise discriminant analysis resembles stepwise regression analysis in that variables are brought into the analysis in order of their ability to contribute to further discrimination between the categories. In Table 20-2 the order of variables entering the analysis is shown. For the seven-variable analysis the Student t-statistic for each variable upon entering the analysis is shown, and the contribution of each variable to the z-score for the difference in discriminant function value between participants and nonparticipants is given. The total standardized difference between groups was 1.297. This means that the cutting plane between groups was roughly 0.65 standard deviations from each group mean. At this level there is an overlap between groups such that roughly 25 percent of each group would be misclassified using the cutting plane value.

Participants were half as often employed as nonparticipants, had incomes lower by one-third, and had six times the mobility problems of nonparticipants. However, they admitted to less trouble in shopping, and felt less isolated and less lonely than nonparticipants. Since these were new participants in the programs, these last differences are factors connected with self-selection into the program and not consequences of the program. To the extent that these programs had secondary objectives of increasing the psychological well-being of participating elderly, it would be important to control for these variables in measuring program success. Those electing to participate already tend to better psychological health than nonparticipants.

A further aspect of the evaluation design was to use the discriminant function to assist in defining groups of persons in areas not served by nutritional centers who would be more like program participants than randomly selected qualifying persons from these areas. The idea was to select an essen-

Table 20-2. Discriminant Analysis for Attendance at Nutritional Centers for the Elderly.

Variable	Entering t-value	Contribution
Personal income	3.46	0.352
Mobility problems	4.00	0.301
Isolation	3.37	0.270
Loneliness	3.06	0.210
Trouble shopping	1.91	0.047
Spend necessity*	1.43	0.066
Unemployed	1.01	0.051
Total z-score difference between means		1.297

*Would spend extra income, if received, on necessities.

tially random sample who would be measured on the variables entering into a discriminant function like the one above. Then values of the discriminant function would be calculated for each person. Those scoring on one side of the cutting point value are more like program participants, and those on the other side are more like nonparticipants. An appropriate comparison group is the first one.

Evaluation and Controlled Experimentation

An ideal in laboratory experimentation in biology is to have a random assignment of experimental material (animals, plants, or whatever) to experimental and control groups. Then, except for presence or absence of treatment, experimental and control groups are alike, and any difference exceeding the limits of chance error can be ascribed to the treatment given the experimental group.

Controlled experimentation is also carried out in matters affecting human populations. Probably the most prevalent example is testing drugs and pharmaceuticals. Here the standard is the *double-blind* experiment, where a placebo (sugar pill) is administered to controls in such a way that neither the subjects nor those administering and making diagnoses know which subjects have received the real treatment and which have received the placebo.

An early large-scale public health experiment was conducted in Scotland in 1930. Called the Lanarkshire milk experiment, its purpose was to evaluate the effects of drinking pasteurized and raw milk on the growth of children. The experimental and control samples involved some 20,000 schoolchildren. W. S. Gosset in 1931 criticized the study for lack of thorough randomization. It appeared that teachers were given some latitude in assigning children to groups to receive milk. Those assigned to receive milk were found to have been shorter and lighter than the no-milk control group at the outset of the experiment. Gossett wrote that 50 pairs of twins with random assignment would have produced more reliable results than the study of 20,000 children. [22]

A very large scale public health experiment was the test of Salk polio vaccine in the United States in 1954. The experiment involved a half a million children in a double-blind experimental versus placebo control approach. The seemingly large numbers were necessary to generate enough cases of polio in the experimental and control groups so that differences in rates could be clearly ascribed to the vaccine. For example, in nearly equal numbers of experimentals and placebos, 16 and 57 cases of paralytic polio were observed. On a null hypothesis of equal expectation, a chi-square of 23.0 is generated, which has a probability of less than 0.005 of being exceeded as a result of chance. The experiment was a success, and the Salk vaccine was introduced nationally but later supplanted by live-virus vaccines. [23]

In the late 1960s Federal laws and regulations increasingly required that Federal domestic programs be evaluated to determine how effectively they were accomplishing their objectives. The Elementary and Secondary Education Act of 1965 required that state agencies determine the effectiveness of payments made under the act, including their relation to measures of educational achievement. Beginning in 1967, evaluation requirements were built into a wide variety of antipoverty program regulations. The concept of demonstration programs employed by the Office of Economic Opportunity was to experiment with new programs long enough to determine their effectiveness before turning successful programs over to traditional agencies. [24]

Daniel Bell regards the social and economic programs of the 1960s as political expressions of the faith in the good effects of government action spawned by the Great Depression and World War II. This faith in the effectiveness of government action was buttressed by optimism about the power of social science to suggest effective avenues for action. [25]

Inferences about causes in observational studies of social programs can be expected to produce controversy. The Civil Rights Act of 1964 had directed the Commissioner of Education to document the existence of differences in educational opportunity occasioned by race, religion, or national origin. The resulting report, *Equality of Educational Opportunity*, commonly called the Coleman Report, reported on tests of 600,000 children and interviews of 60,000 teachers in 4000 schools across the nation. The conclusions reached were: (1) that differences in educational resources available to various minority and ethnic groups *vis-à-vis* other students were small; and (2) the deficiencies in test scores of minority students were more marked in later grades than at entrance into the school systems. Deficiencies at entrance were related to family background and cultural influences of peers. [26]

The report was devastating to liberals because it seemed to say that schools were failing in their mission to equalize opportunities in society, and the lack of relation of educational expenditure to achievement suggested that nothing could be done about it. Methodological criticism of the report centered on the use of cross-sectional analysis to make inferences about dynamic processes and the appropriateness of the method of measuring the net effects of critical variables. [27]

In 1965 the idea of a negative income tax (NIT) as an alternative to traditional welfare programs was proposed independently by Milton Friedman of the University of Chicago and Robert J. Lampman of the University of Wisconsin. [28] The idea of the negative income tax is to provide a support level, g, and a rate at which benefits are reduced by income earned above the support level, r. For example, a support level might be $100 a week and the tax rate might be 30 percent. Weekly payments received by a family under this plan would be

$$P = g - rY,$$

where Y is the family's weekly income. Thus, if the family earned nothing, the

NIT payment would be $100 = g$, and if the family earned \$70 the NIT payment would be

$$P = \$100 - 0.30 \, (\$70) = \$89.$$

If the family earns \$150, which exceeds g, the support level, payment under the plan is

$$P = \$100 - 0.30 \, (\$150) = \$55.$$

Any combination of support level, g, and tax rate, r, has a point at which benefits are reduced to zero, called the "break-even" point. In our example it is

$$\text{BE} = \frac{g}{r} = \frac{\$100}{0.30} = \$333.$$

At issue in the negative income tax proposal was whether individuals receiving benefits would elect to work less than they would have worked in the absence of benefits. Beyond this, the effect of different plans, or values for g and r, on labor force participation was unknown. Championed by David Kershaw within the Office of Economic Opportunity, an experimental program was designed and carried out in several communities in New Jersey from 1968 to 1972. Researchers at the University of Wisconsin planned the study as a controlled experiment. The design called for alternative NIT plans to be applied to randomly selected groups from the target population, and an analysis conducted to determine the effects of different treatments. The analysis was carried out using a multiple regression model in which one set of variables defined the various NIT treatments (plans) and another set represented demographic characteristics that might affect labor force participation. [29]

The major finding was that the reduction in hours worked by male heads of household who received NIT payments was only on the order of 5 percent. The negative income tax welfare idea was never considered seriously by the Johnson administration, but it was considered for a time and then abandoned by the Nixon administration. [30]

There were substantial problems in carrying out the New Jersey NIT experiment, and there were controversies over the methods used and consequently over the conclusions. In any event a later \$60 million study of 4800 families in Denver and Seattle found that NIT plans did reduce substantially the incentives for work. [31]

Ironically, failures of methodology dispose results toward underestimation of program effects. Aaron comments that the major evaluations of education and job programs of the Great Society

> revealed the inadequacies of research methodology more than they documented the ineffectuality of education. These flawed studies, however, marched forth like soldiers to battle, slaying the naively held preconceptions about the effectiveness of education, before falling themselves to criticism and evaluation. [32]

In any event there were some surprising relations between evaluation and program continuance. Among the major poverty programs, Head Start largely failed in its evaluations but continued strongly funded; Job Corps received mixed and positive evaluations but was sharply curtailed; Community Action Programs received generally good evaluations but became the popular symbol of the alleged failure of the War on Poverty. [33]

References

1. Sonja M. McKinley, "The Design and Analysis of the Observational Study—A Review," *Journal of the American Statistical Association*, **70**, No. 351 (September 1975), 503–520.
2. Donald T. Campbell and Julian C. Stanley, *Experimental and Quasi-Experimental Designs for Research*, Rand McNally, Chicago, 1966.
3. Donald B. Rubin, "William G. Cochran's Contributions to the Design, Analysis, and Evaluation of Observational Studies," in *W.G. Cochran's Impact on Statistics* (Poduri S. R. S. Rao and Joseph Sedransk, eds.), Wiley, New York, 1984, pp. 37–69.
4. Leslie Kish, "Some Statistical Problems in Research Design," *American Sociological Review*, **24** (June 1959), 328–338.
5. Edward C. Bryant, "Survey Statistics in Social Program Evaluation," in *Papers in Honor of H.O. Hartley* (H.A. David, ed.), Academic Press, New York, 1978.
6. Thomas D. Cook and Donald T. Campbell, *Quasi-Experimentation: Design and Analysis Issues for Field Settings*, Houghton Mifflin, Boston, 1979.
7. C. West Churchman, *Theory of Experimental Inference*, Macmillan, New York, 1948, pp. 85–116.
8. Kish, op. cit., p. 329.
9. Cook and Campbell, op. cit., p. 23.
10. Campbell and Stanley, op. cit., p. 5.
11. Edward R. Tufte, *Data Analysis for Politics and Policy*, Prentice-Hall, Englewood Cliffs, NJ, 1974, pp. 5–18.
12. Laura Irwin Langbein, *Discovering Whether Programs Work: A Guide to Statistical Methods for Program Evaluation*, Scott, Foresman, Glencoe, IL, 1980.
13. Theodore Colton and Robert C. Buxbaum, "Motor Vehicle Inspection and Motor Vehicle Accident Mortality," in *Statistics and Public Policy* (William B. Fairley and Frederick Mosteller, eds.), Addison–Wesley, Reading, MA, 1977, pp. 131–142.
14. Donald T. Campbell and Robert F. Boruch, "Making the Case for Randomized Assignment to Treatments by Considering the Alternatives: Six Ways in Which Quasi-Experimental Evaluations in Compensatory Education Tend to Underestimate Effects," in *Evaluation and Experiment* (Carl A. Bennett and Arthur A. Lumsdaine, eds.), Academic Press, New York, 1975, pp. 195–275.
15. E. J. Working, "What Do Statistical 'Demand Curves' Show," *Quarterly Journal of Economics*, **4** (1927), 212–235.
16. Mordecai Ezekiel and Karl A. Fox, *Methods of Correlation and Regression Analysis*, Wiley, New York, 1959, p. 420.
17. Sewell Wright, "Correlation and Causation," *Journal of Agricultural Research*, **20** (1921), 557–585.
18. Sewell Wright, "The Method of Path Coefficients," *Annals of Mathematical Statistics*, **5** (1934), 161–215.

19. Sewell Wright, *Evolution and the Genetics of Populations, Vol I Genetic and Biometric Foundations*, University of Chicago Press, Chicago 1968.
20. R. A. Fisher, "The Use of Multiple Measurements in Taxonomic Problems," *Annals of Eugenics*, **7** (1936), 179–188.
21. Kirschner Associates, Inc., *Longitudinal Study Design for Evaluation of the National Nutrition Program for the Elderly*, Contract No. HEW-OS-74-89, Albuquerque, NM, 1974.
22. James W. Tankard, Jr., *The Statistical Pioneers*, Schenkman, Cambridge, MA, 1984, p. 106.
23. Paul Meier, "The Biggest Public Health Experiment Ever: The 1954 Field Trial of the Salk Poliomyelitis Vaccine," in *Statistics: A Guide to the Unknown* (Judith Tanur, ed.), Holden-Day, 1972, pp. 2–13.
24. Henry J. Aaron, *Politics and the Professors: The Great Society in Perspective*, The Brookings Institute, Washington, DC, 1978.
25. Daniel Bell, *The Social Sciences Since the Second World War*, Transaction Books, New Brunswick, NJ, 1982.
26. William W. Cooley and Paul R. Lohnes, *Evaluation Research in Education*, Irvington, New York, 1976, p. 165.
27. Ibid., pp. 218–219.
28. Robert A. Levine, "How and Why the Experiment Came About," in *Work Incentives and Income Guarantees: The New Jersey Negative Income Tax Experiment* (Joseph A. Pechman and Michael A. Timpane, eds.), The Brookings Institute, Washington, DC, 1975, p. 16.
29. Robert Ferber and Werner Z. Hirsch, *Social Experimentation and Economic Policy*, Cambridge University Press, Cambridge, UK, 1982.
30. M. Kenneth Bowler, *The Nixon Guaranteed Income Proposal*, Ballinger, Cambridge, MA, 1974.
31. Robert G. Speigelman and K.E. Yeager, "Overview," *The Journal of Human Resources*, **XV**, No. 4 (1980), 463–479.
32. Aaron, op. cit., pp. 30–32.
33. Sar A. Levitan and Gregory K. Wurzburg, *Evaluating Federal Social Programs*, W. E. Upjohn Institute, Kalamazoo, MI, 1979.

CHAPTER 21

Nonparametrics and Robust Methods

The chapters in this book have traced the origin and development of some of the major ideas and applications of statistics. A large part of this history has to do with inference about the mean of a distribution. In stating a confidence interval or testing a hypothesis about a mean based on sample data, the usual classical technique is to use the Student t-distribution. Indeed, Gosset's discovery of this distribution in 1908 was a major step forward in the history of statistics.

The Student t-distribution refers to the ratio of the difference between a sample mean and the population mean to the estimated standard error of the sample mean in samples of the same size taken from a normal population of values. For an example let us turn to Gosset's famous 1908 paper. [1] He shows the hours of sleep gained by ten patients treated with laevohyoscyamine hydrobromide as reported by others. The calculations needed for an inference about whether the drug would have a positive effect in the parent population are carried out in Table 21-1.

From Table 12-2 we see that 95 percent of the Student t-distribution for nine degrees of freedom lies within 2.262 t-multiples of the mean value of zero. A t-value as large as $+3.68$ would occur less than 50 times in 1000 if the drug had no effect. Gosset goes further and gives odds of 0.9774 to 0.0026, or about 400 to 1 that the drug produces additional sleep. No matter what interpretation is given, it rests on the t-distribution which is derived under an assumption that the sample was drawn from a normally distributed underlying population.

Given a small sample size, as is the case here, the sample observations will not be much help in answering the question of normality. Questions that then arise are at least twofold.

Table 21-1. Calculations Based on Data Given by Gosset.

Patient	Hours sleep X	$X - \bar{X}$	$(X - \bar{X})^2$
1	1.9	−0.43	0.1849
2	0.8	−1.43	2.3409
3	1.1	−1.23	1.5129
4	0.1	−2.23	4.9729
5	−0.1	−2.43	5.9049
6	4.4	2.07	4.2849
7	5.5	3.17	10.0489
8	1.6	−0.73	0.5329
9	4.6	2.27	5.1529
10	3.4	1.07	1.1449
Total	23.3	0.00	36.0810

Mean = 2.33   Standard deviation = $36.08/\sqrt{9} = 2.002$

$$t = \frac{2.33 - 0}{2.002/\sqrt{10}} = \frac{2.33}{0.633} = 3.68$$

(1) How much does it matter if the normality assumption is violated? That is, how different are the stated values (the 95 percent levels, etc.) from those prevailing under various degrees of nonnormality?

(2) Are there methods available which make no assumptions about the shape of the population distribution?

We take up the first question shortly, but let us look at the second question now. The answer is *yes*. Such methods or tests are called *nonparametric*. A nonparametric test that could be applied to the data of Table 21-1 is the *sign-test*. If the drug had no real effect, then the observed effect for each patient would be as likely to be negative as positive. Note that there are nine positive values and one negative value out of ten observations. We can ask what the probability of as many as nine positive signs out of ten is if the true probability of a positive sign were one-half. The answer, from the binomial distribution, is $10/1024 + 1/1024 = 11/1024 = 0.011$. From this small probability we could conclude that the drug had a real effect.

The earliest use of the sign test is attributed to Arbuthnot in 1710. While other early examples could be cited, widespread interest and development of nonparametric tests began in the late 1930s with contributions by Milton Friedman, Harold Hotelling and Margaret Pabst, and Maurice Kendall. [2] It continued in the 1940s and 1950s with contributions by Frank Wilcoxon, William H. Kruskal, W. Allen Wallis, William G. Cochran, Jack Wolfowitz, Quinn McNemar, and others. From the wide variety of nonparametric tests

we will illustrate one more—the Wilcoxon signed rank test for paired observations. We will use again some data provided by Gossett in his 1908 paper.

The Wilcoxon Signed-Rank Test

The data are yields from paired plots planted with regular and with kiln-dried
seed corn. In Table 21-2 we see the sample mean difference of 33.7 pounds per
acre favoring the kiln-dried seed. The Student t-value for the null hypothesis
of no population difference is

$$t = \frac{33.7 - 0}{66.17/\sqrt{11}} = 1.69,$$

which is smaller than the 5 percent cut-off value (two-tailed) in Table 12-2.
The t-value is at the 93.4 percentile of t, showing that only 7 percent of sample
differences would exceed 33.7 in sampling from a population in which there
was no difference in yields. Gosset gives odds of 14 to 1 (93.4 to 6.6) that
kiln-dried corn gives the higher yield. In the final column the absolute differences
are ranked from small to large and then the sign of the difference attached.
The nonparametric test for difference is based on the total of the signed ranks,
a sum we will call V.

If each pair of observations came from the same population, then the sign
of any difference would be as likely to be plus as minus, and the expected sum
of the signed differences would be zero. When the number of pairs, n, is 10
or more, the probability distribution of V under the hypothesis of sampling
from identical populations is approximately normal with standard deviation

Table 21-2. Corn Yield (lbs per Acre) from Paired Plots of
Regular and Kiln-Dried Seed.

Pair	Regular	Kiln-dried	Difference	Signed-rank	
1	1903	2009	+ 106	+ 10	
2	1935	1915	− 20	− 1	
3	1910	2011	+ 101	+ 9	
4	2496	2463	− 33	− 3	
5	2108	2180	+ 72	+ 8	
6	1961	1925	− 36	− 4	
7	2060	2122	+ 62	+ 6	
8	1444	1482	+ 38	+ 5	
9	1612	1542	− 70	− 7	
10	1316	1443	+ 127	+ 11	
11	1511	1535	+ 24	+ 2	
Average	1841.5	1875.2	+ 33.7	+ 36	(total)

equal to

$$\sigma_V = \sqrt{\frac{n(n + 1)(2n + 1)}{6}}.$$

The hypothesis of no difference can be tested by the standard normal deviate

$$z = \frac{V}{\sigma_V}.$$

For the data of Table 21-2

$$\sigma_V = \sqrt{\frac{11(12)(23)}{6}} = 22.5,$$

and

$$z = \frac{36}{22.5} = 1.69,$$

which is approximately at the 94.5 percentile of the standard normal deviate.

The sign test and the signed-rank test are nonparametric alternatives to the *t*-test for mean difference. They can be used when one has doubts about the validity of the normality assumption required for the Student *t*-test. Nonparametric techniques are rated by how efficient they are in comparison to the alternative parametric method when the assumptions underlying the parametric method are met. The criterion of efficiency is *power* to reject the null hypothesis when it is false. For example, the sign test has an efficiency of 0.637 compared to the Student *t*-test. This means that the Student *t*-test achieves the same power as the sign test with a sample 0.637 times as large. The efficiency of the Wilcoxon signed-rank test in comparison to Student's *t* is 0.955. Thus, use of the signed-rank test instead of Student's *t* would be valid whether or not the normal assumption was met, an in those cases where it was met (and *t* valid), very little efficiency would be lost by using the Wilcoxon test.

Robustness and Student's *t*

The effects of nonnormality on the validity of Student's *t*-distribution have been studied. For example in a 1960 simulation study, Boneau found that of 1000 *t*-ratios generated by samples of size 10 from an exponential distribution, 3.1 percent fell in the extreme 5 percent of the tabled distribution and 0.3 percent fell in the extreme 1 percent. From a rectangular distribution, 5.1 percent of 1000 *t*-values from samples of size 10 fell in the extreme tabled 5 percent and 1.0 percent fell in the extreme 1 percent. The tabled *t*-levels appear conservative in the case of the exponential and quite exact in the case of the rectangular with these sample sizes. [3] The general conclusion seems to be that the Student *t*-distribution is quite *robust* in the face of some nonnormal parent distributions for quite modest sample sizes. [4]

The type of robustness just mentioned has come to be called *validity* or *criterion* robustness. By this is meant the approximate correctness of a procedure in the face of violations of assumptions required for its strict correctness.

In the last decade statisticians have been concerned with another kind of robustness called robustness of *efficiency*. To illustrate this concept and the techniques developed to insure this kind of robustness, we use Darwin's data on differences between heights of pairs of *Zea mays* (cross versus self-fertilized) grown in the same pots. The data (in eighths inches) were

Pairs:	1	2	3	4	5	6	7	8	9	10	11	12	13	14	15
	49	−67	8	16	6	23	28	41	14	29	56	24	75	61	−48

The Student *t*-statistic for the mean difference of 20.93 was 2.148 compared with a critical value of 2.145 for significance at an 0.05 risk of erroneously concluding that a real difference existed. The null hypothesis of no difference is (barely) rejected.

Boxplots were introduced in Chapter 3 for comparing distributions. They are also used for comparing a set of data with the normal distribution form. Figure 21-1 shows how the boxplot is used for this purpose. First, the box extends from the first to the third quartile, encompassing the middle 50

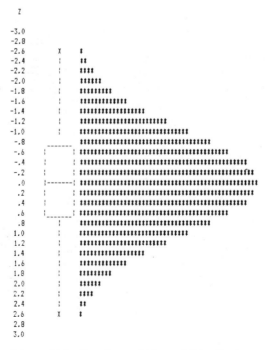

Figure 21-1. Boxplot and Normal Distribution.

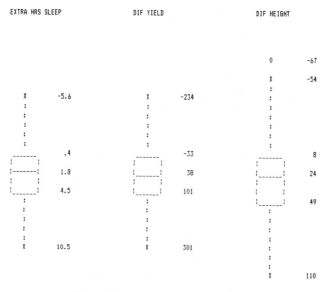

Figure 21-2. Boxplots and Outliers.

percent of values in a data set. In the normal distribution the limits of the box are at plus and minus two-thirds of a standard deviation unit. Then, whiskers are drawn at 1.5 times the box length from the ends of the box. These extremes are then at about 2.7 standard units from the mean. In a normal distribution, the probability of observations beyond *one* of these limits is 0.0035. The percentage of observations *within* these limits in a normal distribution is 99.3.

Figure 21-2 shows the data sets for extra hours sleep and differences in corn yield from kiln-dried versus regular seed that we have already seen in this chapter. Also shown in boxplot form is Darwin's data for differences in plant heights. First, we note the symmetry of the boxes for the first two data sets, indicating that the data are symmetrical. Darwin's data, however, are not symmetrical. The median height difference is 24, which is at a distance of 16 from the first quartile, while the third quartile is 25 units away from the median. Further, we find an observation outside of the whiskers. If the 15 observations were drawn from a normal distribution, the probability of one or more observations beyond these limits is $1 - (0.993 \exp 15) = 0.100$. What we have is evidence of skewness and excess observations in the tails of the distribution.

Robust Estimation

Whatever the reason for an outlier like the one just identified, it calls into question the normality of the underlying population. Presence of extreme observations that are improbable given a normal underlying distribution tend to inflate the sample standard deviation. The presence of long-tailed distributions was recognized by Gosset in a 1927 paper entitled "Errors of Routine Analysis." [5]

Beginning in the 1960s the computer has been available to simulate repeated sampling from any desired population form. Statisticians began to study alternatives to the arithmetic mean for estimating the central location of a population. The estimators proposed were generally ones that would be less affected by extreme values from sample to sample. In 1972 the results of a massive study under the leadership of John Tukey at Princeton were published. [6] The study compared the behavior of some 65 estimators in 35 sampling situations using samples of sizes 5, 10, 20, and 40. [7]

One class of estimators is known as *trimmed means*. In a trimmed mean a certain fraction of observations at both extremes of the data are left out (trimmed) and the mean of the remaining observations are calculated. For example a 25 percent trimmed mean would be the average of the middle half of the observations. The best amount of trimming depends on how much heaviness in the tails of a distribution one wants to protect against, with 10–25 percent trimming recommended for light to moderate excess tails. [8]

For the standard error of a trimmed mean one calculates what is called the *Winsorized* standard deviation. This is a standard deviation calculated after the trimmed observations are replaced by the most extreme of the observations not trimmed. Table 21-3 contains the data and results for 95 percent confidence intervals based on various degrees of trimming for Darwins's differences in plant heights. We see that the influence of the outlying observation is so reduced that we are able to conclude with increased confidence that a real difference in favor of cross-fertilization existed.

G. E. P. Box and G. C. Tiao used Darwin's data to investigate some aspects of robustness. They found that if the sample was assumed to come from a rectangular distribution, the significance level of the data was hardly changed from the level assuming a normal distribution. This represents criterion robustness. They also found that the probability of a difference favoring cross-fertilization depends on the shape of the prior distribution (with mean of zero) assumed for the mean difference. [9] This is lack of efficiency robustness. The results varied when different assumptions were used for the prior distribution. The role of the trimmed mean is to provide a method whose efficiency is resistant to departures from normality of the kind described first.

Steven M. Stigler studied 24 data sets derived from classic scientific experiments made in the eighteenth and nineteenth centuries to determine constants which are now much more accurately known. The sets came from eighteenth century observations made to determine the distance from the earth to the sun, from experiments in 1879–1882 by Michelson and Newcomb to determine the speed of light, and from a 1798 study to determine the mean density of the earth. Grouping the data into samples of around 20, Stigler made estimates of true means using eleven estimators, which included 10, 15, and 25

Table 21-3. Trimmed Means, Winsorized Standard
Deviations, and Resulting t-Statistics for Darwin's Data.

X	Original	\multicolumn{4}{c}{Squared deviations = $(X - \bar{X})^2$}
		Trim 1	Trim 2	Trim 3
-67	7732.27	5117.75	429.62	316.05
-48	4751.80	5117.75	429.62	316.05
6	223.00	307.60	429.62	316.05
8	167.27	241.44	350.71	316.05
14	48.07	90.98	161.98	138.72
16	24.34	56.83	115.07	95.60
23	4.27	0.29	13.89	7.72
24	9.40	0.21	7.44	3.16
28	49.94	19.91	1.62	4.94
29	65.07	29.83	5.17	10.38
41	402.67	304.91	203.71	231.72
49	787.74	648.29	496.07	539.27
56	1229.67	1053.75	856.89	787.74
60	1526.20	1329.44	856.89	787.74
75	2923.20	1329.44	856.89	787.74
$\bar{X}$	20.93	23.54	26.73	25.78
$\sum(X - \bar{X})^2$	19944.93	15648.43	5215.21	4658.92
d.f.	14.00	12.00	10.00	8.00
Std. dev.	37.74	34.69	21.77	22.75
Std. error	9.75	9.62	6.57	7.58
t-value	2.15	3.38	4.07	3.39

percent trimmed means. Knowing approximate true values from modern
studies, he could analyze the performance of the alternative estimators. Based
on relative error from the true values, the 10 and 15 percent trimmed means
performed best. [10] The data appeared to have only moderately heavy tails
and modest positive skewness compared to normal distributions. The same
can be said of Darwin's data.

The Jacknife

A technique that is useful in utilizing nonstandard robust estimators is *the
jacknife*. When the formula for the standard error of an estimator is not
known, the *jacknife* can be used to estimate the standard error. In the jacknife
process, each observation is dropped from the sample in turn, and the estimate
in question calculated. The reason for dropping the observation is to provide a
basis for calculating values of the estimate for different samples. Each sample
has only one observation that is different from other samples, so that the

variance observed is $1/(n - 1)$ times what the variance would be for independent samples of size n. The jacknife was introduced in 1956 by the British
statistician, M. L. Quenouille. [11].

In Table 21-4 we calculate the *trimean* for Darwin's data. The trimean is a
weighted mean of the first and third quartiles and the median, with the median
being given twice the weight of either quartile. It is one of the battery of robust
estimators designed to protect against the effects of outliers and heavier than
normal tails.

The sample trimean turns out to be 26.25 (eighths inches). To conduct a
significance test or establish a confidence interval for this mean requires that
we know its standard error. In the table we show the sample trimean for all
possible samples of size $n - 1 = 14$. There are 15 of them. Then we establish
the estimated standard error from the sum of squared deviations among these
different estimates by the jacknife formula [12]

$$SE = \sqrt{\frac{(n - 1)\sum(V - \bar{V})^2}{n}},$$

where V stands for the estimates with one observation deleted.

$$SE = \sqrt{\frac{14(47.93)}{15}} = 6.69.$$

Table 21-4. Trimeans for Jacknife Estimate of Standard Error.

Observations i	Difference X	Values with $X(i)$ deleted			
		Q1	Median	Q3	Trimean
1	−67	14	26	49	28.75
2	−48	14	26	49	28.75
3	6	14	26	49	28.75
4	8	14	26	49	28.75
5	14	8	26	49	27.25
6	16	8	26	49	27.25
7	23	8	26	49	27.25
8	24	8	25.5	49	27.00
9	28	8	23.5	49	26.00
10	29	8	23.5	49	26.00
11	41	8	23.5	49	26.00
12	49	8	23.5	41	24.00
13	56	8	23.5	41	24.00
14	61	8	23.5	41	24.00
15	75	8	23.5	41	24.00
All 15 Observations		8	24	49	26.25

The observed trimean then has a Student t-statistic for a test that the population trimean is zero of

$$t = \frac{26.25 - 0}{6.69} = 3.96.$$

For 14 degrees of freedom, this value is beyond the 0.01 two-tailed level of Student's t-multiple.

We have seen that two different robust techniques, the trimmed mean with Winsorized standard deviation and the trimean with standard error estimated by the jacknife give significance levels for Darwin's data of the order of 0.01. The simplest nonparametric test, the sign test would ask what the probability is for 13 or more (out of 15) differences in the same direction when the probability of one difference in either direction was 0.50. The answer from the binomial distribution is 0.0074, about the same degree of significance as the two robust estimators. We can conclude that the marginal significance of the traditional t-statistic for the mean difference reflects the lack of robustness of efficiency of the traditional procedure.

Exploratory Data Analysis

Exploratory data analysis is an outlook with accompanying techniques that has gained popularity in the last 15 years. John W. Tukey of Princeton University and the Bell Laboratories was the originator and proponent of much of this approach. [13] The methods fit in well with the use of robust inferential techniques because they emphasize recognition of underlying distribution shapes through the use of boxplots and scatter diagrams. Exploratory data analysis, or EDA, is sometimes contrasted with traditional inferential methods by emphasizing that EDA is an *exploratory* mode of analysis while traditional methods are confirmatory in nature. [14] Traditional methods are well suited to testing theories (relations between variables) that have already been formulated, while EDA is presented as a set of methods for discovering and summarizing a relationship in the first instance. In this sense, EDA is part of *descriptive* statistics. Our example will be a time series analysis of quarterly factory shipment of dryers in the United States. The analysis will be a mixture of conventional descriptive statistics and exploratory data analysis.

Table 21-5(a) presents the original quarterly dryer shipments from 1975 through 1983. The data are arranged in a table where the quarters are the rows and the years are presented in nine columns. A basic way to analyze such a table is shown in part (b). There, the grand mean, 803, is repeated from part (a). Then, the column and row means from part (a) are reduced to deviations from the grand mean. The set of row deviations emphasizes a seasonal, or

Table 21-5. Row and Column Analysis of Dryer Shipments.

(a) Original Data (thousands)

Quarter	1975	1976	1977	1978	1979	1980	1981	1982	1983	Mean
1st	608	840	876	925	925	921	801	715	794	823
2nd	640	696	779	855	804	634	713	623	782	725
3rd	885	826	952	913	902	769	790	685	832	839
4th	737	810	946	928	872	855	666	705	886	823
Mean	718	793	888	905	876	795	743	682	824	803

(b) Grand Mean, Row and Column Deviations, and Residuals

Quarter	1975	1976	1977	1978	1979	1980	1981	1982	1983	Mean
1st	−130	27	−33	−1	29	106	38	13	−50	20
2nd	0	−20	−32	27	6	−83	48	18	36	−77
3rd	131	−4	27	−29	−11	−63	11	−34	−28	37
4th	−1	−3	37	2	−24	40	−97	3	42	20
Mean	−85	−10	86	103	73	−8	−60	−121	21	803

(c) Analysis of Row (Seasonal) Deviations

Quarter	1975	1976	1977	1978	1979	1980	1981	1982	1983	Mean	Median	Adjusted
1st	−110	47	−12	20	49	126	59	33	−30	20	33	31
2nd	−78	−97	−109	−50	−72	−161	−30	−59	−42	−77	−72	−76
3rd	168	33	64	8	26	−26	48	3	9	37	26	24
4th	20	17	58	23	−4	60	−77	23	63	20	23	21
Mean											10	0

(d) Adjusted Residuals

Quarter	1975	1976	1977	1978	1979	1980	1981	1982	1983	Mean
1st	−141	16	−44	−12	18	95	27	2	−61	31
2nd	−1	−21	−33	26	5	−84	47	17	35	−76
3rd	144	9	40	−16	2	−50	24	−21	−15	24
4th	−2	−4	36	1	−25	39	−98	2	41	21
Mean	−85	−10	86	103	73	−8	−60	−121	21	803

within year pattern of variation in the data. Second quarter shipments average 77,000 less than average quarterly shipments over the period, and this is balanced off by above average shipments in the other three-quarters. The column means show an among year pattern in which average quarterly sales begin below the overall average, rise above it in 1977 through 1979, and fall below the average in 1980 and 1982 only to rise above the average quarterly rate in 1983.

The body of part (b) shows what are called residuals. Each observation (quarterly shipments) is viewed as being made up of a common level plus a year (column) effect plus a seasonal (row) effect plus a residual (unexplained) variation.

$$X = \bar{X} + d(\text{column}) + d(\text{row}) + r.$$

For example, fourth quarter 1978 shipments were 928,000. This is made up of

$$d(\text{column}) = 905 - 803 = 103,$$
$$d(\text{row})\quad\; = 823 - 803 = 20,$$
$$r = X - \bar{X} - d(\text{column}) - d(\text{row})$$
$$= 928 - 803 - 103 - 20 = 2.$$

Thus,

$$928 = \bar{X} + d(\text{column}) + d(\text{row}) + r$$
$$= 803 + 103 + 20 + 2.$$

In part (c) of Table 21-5 we pay special attention to the deviations of original values from the column means with the objective of obtaining the best measure of the seasonal pattern in the data. The averages in the mean column are the same as from part (b), but we also show the median of the deviations in each row. It is argued that the median is more *resistant* to (less affected by) extremes or outliers in a set of data. In fact, as the boxplots in Figure 21-3 show, we have outliers in the third and fourth quarter sets of 168 and -77, respectively. The third quarter outlier has an especially large influence on the mean. The mean is 37 while the median is 26. Relying on the column of medians as the better seasonal measure, we then adjust them so their total balances to zero. These are our final seasonal factors.

In part (d) the residuals from part (b) are adjusted for the changes made to arrive at the seasonal factors. For example, because 11 was added to the average deviation for the first quarter, 11 is subtracted from all the residuals. The column deviations for the years are unchanged. Part (d) replaces part (b) as the data summary.

Figure 21-4 shows some further descriptions of the dryer shipments data that are informative. First, the original data are adjusted for seasonal variation. This means that the deviations associated with the quarters from Table 21-5 are removed from the data. That is,

$$\text{First quarter: } X(\text{adj}) = X - 31,$$
$$\text{Second quarter: } X(\text{adj}) = X - (-76),$$
$$\text{Third quarter: } X(\text{adj}) = X - 24,$$
$$\text{Fourth quarter: } X(\text{adj}) = X - 21.$$

The result, plotted in Figure 21-4, is a series that emphasizes the nonseasonal variations in the data. These are the longer and intermediate term variations associated with underlying growth and with business cycles, as well as the very short-term changes that are termed irregular variation.

Often it is the cyclical changes that analysts wish to bring out. The longer-term changes in a series can be separated from the shorter-term changes by the use of *moving averages*. A variety of moving average procedures are available. [15] We selected a running median smoother as illustrative. We used a 4-term

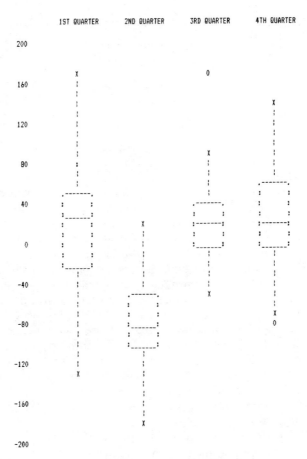

Figure 21-3. Boxplots for Seasonal Deviations.

median smoother followed by a 2-term smoother. The second operation further smooths and centers the 4-term smoothed series. The process, including the preceding adjustment for seasonal variation, is illustrated below for the beginning of the dryer series.

Original data	608	640	885	737	840	696	826	810
Seasonal adjustment	−31	76	−24	−21	−31	76	−24	−21
Adjusted series	577	716	861	716	809	772	802	789
4-term median			716	763	744	759	796	796
Moving average			739	753	752	777	796	810
Adjusted series − Moving average			122	−37	58	−5	7	−21

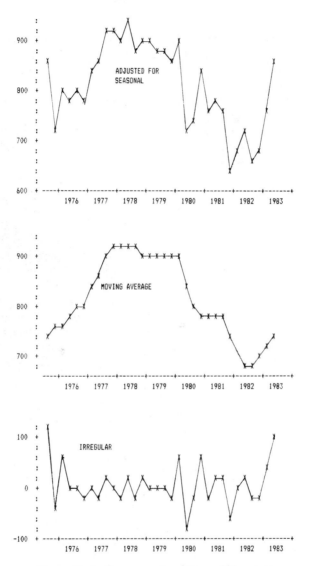

Figure 21-4. Components of Dryer Shipments.

The median of the first four terms in the adjusted series is 716 since both the second and third ranking observations have that value. For the second four values (716, 861, 716, 809) the median is (716 + 809)/2 = 763. The moving average is obtained by averaging the median of the first four and the median of the second four values, (716 + 763)/2 = 739. The process continues in this fashion. A 4-term median rather than a 4-term mean is used because the median is more resistant to outliers. The 2-term average (which is both median and mean) further smooths and also centers the final result. An average of the first four terms is properly centered between the second and third terms and an average of the second four terms is properly centered between the third and fourth terms. The average of the first two medians is thus properly centered at the third time period.

The moving average, plotted in the second panel of Figure 21-4, does a better job of showing the long-term trend and cyclical variation in the dryer shipments because the very short-term irregular variation has been smoothed out. In EDA one speaks of the *smooth* and the *rough*. The longer-term variations are the smooth, which have been preserved in the moving average. The irregular variation, shown in the final panel, is obtained by subtracting the moving average from the original adjusted series. This rough variation is much like the residual variation in our earlier analysis.

The pattern of the moving average follows the annual pattern of the column means in the earlier tabular analysis. Each emphasizes that cyclical variation is a major element of variability in dryer shipments. To predict dryer shipments one would have to understand and predict this variation.

Our final analysis is directed toward explaining the changes in the column means of the original table. In Table 21-6 we show these average quarterly shipment figures along with annual figures for residential investment (home building) in the United States. The year-to-year changes in dryer shipments and residential investment are plotted in Figure 21-5. There we see that the two are very closely related. The regression of the changes, or *first differences*, is

$$\text{Est. Dif. } Y = 1.702 + 8.798 \text{ Dif } X,$$
$$r^2 = 0.96.$$

Table 21-6. Data for Difference Regression of Dryer Shipments on Residential Investment in the United States.

Year	Average quarterly shipments (thousands)	Residential investment (billions)	Annual change in shipments	Annual change in investment
1975	718	$42.2		
1976	793	51.2	75	9.0
1977	888	60.7	95	9.5
1978	905	62.4	17	1.7
1979	876	59.1	−29	−3.3
1980	795	47.1	−81	−12.0
1981	743	44.7	−52	−2.4
1982	682	37.8	61	−6.9
1983	824	52.7	142	14.9

*SOURCE: *Survey of Current Business* and *Economic Indicators*.

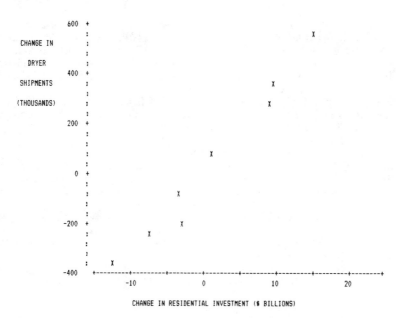

Figure 21-5. Annual Changes in Dryer Shipments Versus Changes in Residential Investment.

The regression says that when there is no change in residential investment, dryer shipments are expected to increase by 1.702 thousand per quarter. For every billion dollar increase in the change in residential investment, dryer shipments are expected to change by an additional 8.798 thousand per quarter. The relationship over the years studied has been very consistent.

We see that if we are able to predict changes in residential investment, we have an excellent clue to changes in sales of dryers. The industrial forecaster's efforts can be concentrated on economic factors influencing the housing market.

References

1. William S. Gosset, "The Probable Error of a Mean," *Biometrika*, **6** (1908), 1–25.
2. E. H. Lehman, *Nonparametrics*, Holden-Day, San Francisco, 1975.
3. C. A. Boneau, "The Effects of Violations of Assumptions Underlying the *t*-Test," *Psychological Bulletin*, **57** (1960), 49–64.
4. John Gaito, "Non-parametric Methods in Psychological Research," *Psychological Reports*, **5** (1959), 115–125.
5. William S. Gosset, "Errors of Routine Analysis," *Biometrika*, **19** (1927), 151–164.
6. D. F. Andrews, P. J. Bickel, F. R. Hampel, P. J. Huber, W. H. Rogers, and J. W. Tukey, *Robust Estimates of Location: Survey and Advances*, Princeton University Press, Princeton, NJ, 1972.

7. Colin Goodall, "*M*-Estimators of Location: An Outline of the Theory," in *Understanding Robust and Exploratory Data Analysis* (David C. Hoaglin, Frederick Mosteller, and John W. Tukey, eds.), Wiley, New York, 1983, pp. 339–403.

8. James L. Rosenberger and Miriam Gasko, "Comparing Location Estimators: Trimmed Means, Medians, and Trimean," in Hoaglin, Mosteller, and Tukey, op. cit., 297–338, p. 329.

9. G. E. P. Box, "Robustness in the Strategy of Scientific Model Building," in *Robustness in Statistics* (Robert W. Launer and Graham N. Wilkinson, eds.), Academic Press, New York, 1979, pp. 201–236.

10. Stephen M. Stigler, "Do Robust Estimators Work with *Real* Data?," *The Annals of Statistics*, **5**, No. 6 (1977), 1055–1098.

11. M. H. Quenouille, "Notes on Bias in Estimation," *Biometrika*, **43** (1956), 353–360.

12. Bradley Efrom and Gail Gong, "A Liesurely Look at the Bootstrap, the Jacknife, and Cross-Validation," *The American Statistician*, **37**, No. 1 (February 1983), 36–48.

13. John W. Tukey, *Exploratory Data Analysis*, Addison-Wesley, Reading, MA, 1977.

14. Frederick Hartwig and Brian E. Dearing, *Exploratory Data Analysis*, Sage Publications, Beverly Hills, CA, 1979, p. 10.

15. Paul F. Velleman and David C. Hoaglin, *Applications, Basics, and Computing of Exploratory Data Analysis*, Duxbury Press, Boston, MA, 1981, Ch. 6.

CHAPTER 22

Statistics—A Summing Up

Our journey that began with William Petty's political arithmetic and Blaise Pascal's correspondence on gaming is at an end. The tour has provided an introduction to the principles and practice of economic and social statistics. The journey has taken us back in time and we have viewed the principal developments in the mirror of history.

Petty and Pascal serve to remind us of the double origins of statistics in mathematics and political economy. Remember it is barely 100 years since these original strands were joined in the work of Galton and K. Pearson. In that 100 years the world has benefited from the advances begun by Gosset, R. A. Fisher, Shewhart, Neyman, Wald, Gallup, and others. We have met a number of these statisticians in these pages.

Table 22-1 presents an historical array of personalities. It begins with Pascal and Petty. The separate branches emphasize the dual foundations of mathematics and political arithmetic. Who goes where is partly arbitrary. For example, Edmund Halley was a competent mathematician and astronomer, and a personal friend of De Moivre. But we place him on the political arithmetic side because he constructed an early table of life expectancies. Quetelet could be placed on either side. Indeed he was the first to combine description of social phenomena with mathematics.

Arbuthnot was a friend and collaborator of Jonathan Swift, author of *Gulliver's Travels*, and of the poet and essayist Alexander Pope. He was a physician, a serious scholar of science and humanities, and a political satirist. He created the figure of John Bull to represent England, along with Lord Strutt (Charles II of Spain), Nicholas Frog (Holland), and Louis Baboon (Louis XIV of France). He extended the "argument from design" for the existence of God with the observation that 82 successive years of excess of male over female births in London had a probability of 1/2 to the eighty-

Table 22-1. Chronology.

B. Pascal 1623–1662	Galileo 1564–1642	J. Graunt 1620–1674
P. Fermat 1601–1655		Wm Petty 1623–1687
J. Bernoulli 1654–1705	John Locke 1632–1704	J. Arbuthnot 1667–1735
A. deMoivre 1667–1754	Isaac Newton 1642–1727	E. Halley 1656–1742
D. Bernoulli 1700–1782		J. Süssmilch 1707–1767
T. Bayes 1702–1761	David Hume 1711–1776	G. Achenwall 1719–1772
P. Laplace 1749–1827		B. Franklin 1706–1790
C. F. Gauss 1777–1853	T. R. Malthus 1766–1834	J. Sinclair 1754–1835
S. Poisson 1781–1840		L. Shattuck 1793–1859
A. Quetelet 1796–1874		E. Jarvis 1803–1884
	J. S. Mill 1806–1873	W. S. Jevons 1835–1882
F. Nightingale 1820–1910	Charles Darwin 1809–1882	F. Walker 1840–1897
C. Booth 1840–1916	Francis Galton 1822–1911	C. D. Wright 1840–1909
F. Y. Edgeworth 1845–1926		

Quality control	Sampling and surveys	Experiment and observation	Decision-making	Nonparametrics and exp. data analysis	Factor analysis and mental testing	Econometrics	Time series and forecasting
W. Shewhart 1891–1967			Karl Pearson 1857–1936		C. E. Spearman 1863–1945	H. Moore 1869–1958	A. L. Bowley 1869–1957
			W. S. Gosset 1876–1937		G. Thomson 1881–1955	H. Schultz 1893–1938	G. Yule 1871–1951
H. F. Dodge 1893–1976	F. Yates 1902–	H. Snedecor 1881–1974	R. A. Fisher 1890–1962	H. Hotelling 1895–1973	L. L. Thurstone 1887–1955	M. Ezekiel 1899–1974	S. Kuznets 1900–
LHC Tippett 1902–	P. Mahalanobis 1893–1972	G. Cox 1900–1978	J. Neyman 1894–1981	F. Wilcoxon 1892–1965	K. Holzinger 1892–	R. Frisch 1895–1973	M. Kendall 1907–1983
W. E. Deming 1900–	G. Gallup 1901–1984	W. G. Cochran 1909–1980	A. Wald 1902–1950	M. Friedman 1912–	S. S. Wilks 1906–1964	J. Tinbergen 1903–	J. Shiskin 1912–1978
J. M. Juran 1904–	M. Hansen 1910–	D. T. Campbell 1916–	L. J. Savage 1917–1971	J. Tukey 1919–	H. Solomon 1919–	L. Klein 1920–	G. E. P. Box 1919–

second power under a "chance" hypotheses. Karl Pearson later pointed out that "the idea of an omnipresent activating deity, who maintains mean statistical values" persisted in statistical thought from Süssmilch through Quetelet to Florence Nightingale. [1]

Our arrangement shows Karl Pearson at the junction of the two branches. Florence Nightingale was an intellectual admirer of Quetelet and a friend of Francis Galton. Pearson drew examples for his early lectures from the work of Charles Booth. Lemuel Shattuck, the first president of the American Statisticial Association, corresponded with Quetelet. The economist Jevons is included because of his interest in statistical description of economic series. Recall that Walker and Wright were instrumental in the design of late nineteenth-century censuses and labor statistics programs. Wright was well known internationally.

In the middle stem we show some people who influenced scientific and political thought—Galileo, Locke, Newton, Hume, Malthus, and Mill. The main-stem chronology after Pearson shows some of the major contributors to the foundations of statistical inference and decision-making. On the ancillary branches some of the major contributors to several areas of application are shown.

Maurice Kendall picks 1890 as the beginning date for modern statistics. In that year Galton was 68, F. Y. Edgeworth was 45, and Karl Pearson was lecturing at University College and thinking about what ultimately became *The Grammar of Science*. G. Udney Yule, barely 20 years of age, was a student at University College who would return in 3 years as Pearson's assistant. [2]

We are now in the fourth academic generation since Pearson, reckoning generations at around 25 years. Our table identifies some key figures from the first two of these generations. We see Yule and Gosset, Pearson's students, contributing to time series and main-stem developments. Spearman, a contemporary of Pearson, is shown as first in a sequence centered around factor analysis and mental testing. However, Galton is recognized as the father of this field. [3] Shewhart begins the list in quality control, and Harold Dodge, his contemporary, led in the development of sampling inspection plans for control of incoming quality.

With Fisher's generation we see further American influence. Moore and Schultz, the pioneers in econometrics, were American, as well L. L. Thurstone and K. Holzinger, who introduced ideas in multiple factor analysis. Harold Hotelling took his Ph.D. in mathematics from Princeton and made contributions to both mathematical economics and statistics during his career. Paul Samuelson rates his contributions to economics along with those of Schultz, Tinbergen, and Frisch. [4] Lawrence Klein, the leading force behind the Wharton econometric model of the United States, received the Nobel Prize in economics for his work in 1980.

G. Udney Yule resigned as Pearson's demonstrator after 6 years to take a secretarial post. But he continued to give lectures that ultimately became his text, *Introduction to the Theory of Statistics*, first published in 1911. By this time Yule had extended and clarified the theory of partial correlation and linear regression for any number of variables. Yule's later work in 1926–1927 focused on correlation of time series. He laid the foundations for the study of autoregressive patterns in time series, which was extended by M. G. Kendall and others. [5]

University centers of statistical development in the United States have included Columbia, North Carolina, California–Berkeley, and Iowa State. George W. Snedecor built the statistics area at Iowa State during the 1920s, and brought R. A. Fisher to the United States twice in the 1930s. The connection between Iowa State and the Department of Agriculture through Henry A. Wallace was mentioned in Chapter 13. Mordecai Ezekiel developed applied regression studies with the Bureau of Agricultural Economics during the 1920s, and joined Wallace as economic advisor to the Secretary of Agriculture in 1933.

Another visitor to the United States (in 1938) was L. H. C. Tippett from the British Cotton Industry Research Association. He gave a series of lectures in quality control at the Massachusetts Institute of Technology. His book, *Technological Applications of Statistics*, was an outgrowth of a second set of lectures given at MIT after the war. Jerzy Neyman visited the Department of Agriculture in 1937 at the instigation of Deming and S. S. Wilks. Neyman returned in 1938 to remain at Berkeley. The work of Deming and J. M. Juran in statistical quality control in the United States and Japan was mentioned in Chapter 16. Deming's contributions to sample survey design are also noteworthy.

Another who came to stay was W. G. Cochran, who joined Iowa State in 1939 after studying at Cambridge University and serving at Rothamsted Experimental Station in England. His major texts were the revision of Snedecor's *Statistical Methods* and *Experimental Designs* with Gertrude Cox. Cox was Snedecor's first M.A. student at Iowa State (1929) who went on to organize the Department of Experimental Statistics at North Carolina State in Raleigh in 1941 and the Institute of Statistics at the University of North Carolina in Chapel Hill in 1949. Cochran assisted in this last effort before going on to Johns Hopkins and Harvard Universities. [6] D. T. Campbell is recognized for his contribution, with Julian Stanley, to the analysis of quasi-experimental designs in observational studies.

Hotelling organized the Statistical Research Group (SRG) at Columbia to assist the military in World War II. In Chapter 16 the members of the group were listed. Included were L. J. Savage, Abramham Wald, and Milton Friedman. Hotelling's contributions to statistics include principal components analysis and multivariate extensions of difference tests and correlation (Hotelling's T-squared and canonical correlation). After World War II he moved to the University of North Carolina.

The infusion of economics on mainstream statistical thought via its role in decision-making could stem from Hotelling and the Statistical Research Group at Columbia. While Abraham Wald, who formalized decision theory, and L. J. Savage, who moved the mainstream from decision theory to Bayesian methods, were not economists, Milton Friedman and George Stigler of the group were, and one could count Hotelling also. The mixture was surely there. Friedman and Wallis contributed an important nonparametric test, another area which Hotelling pioneered. Friedman and Savage collaborated on the S-shaped utility curve paper, and yet another SRG member, Frederick Mosteller, produced a key paper with Philip Nogee on measurement of utility. Herbert Solomon, also one of the group, concentrated his career on multivariate classification and clustering techniques.

Frank Wilcoxon received a Ph.D. in chemistry in 1924 from Cornell University and worked in agricultural and industrial chemistry culminating in 13 years with the American Cyanamid Company. His interest in statistics began early in his industrial career with a study group working through Fisher's *Statistical Methods for Research Workers* in 1925. In 1945 he published two papers that introduced the rank-sum test for two samples and the signed-rank test for paired sample observations. A few years after his retirement from industry he joined the faculty of Florida State University. [7] John Tukey's great contributions are exploratory data analysis (EDA) and robust methods, which we touched on in Chapter 21.

Yule's text went to ten editions by 1935, by which time Yule felt that a co-author should be brought in. The co-author turned out to be Maurice Kendall, a civil servant who had developed a strong interest in statistics. By 1950 Yule and Kendall was in its fourteenth edition. [8] In the meantime Kendall had completed a project begun by a group of five British statisticians to summarize the state-of-the-art in statistics. The result was Kendall's *The Advanced Theory of Statistics*, Vol. 1 (1943) and Vol. 2 (1946). He contributed to the theory of time series analysis during the 1940s, and in 1949 accepted a chair at the London School of Economics. In the 1950s his interests extended to the areas of multivariate analysis and factor analysis. [9]

A. L. Bowley and Simon Kuznets contributed to the development of national income measures and series in Great Britain and the United States, respectively. Julius Shiskin headed the business cycle statistics program of the Census Bureau and later in his career was the Comissioner of the Bureau of Labor Statistics. G. E. P. Box, who early in his career developed the concept of evolutionary operation (EVOP) in industrial testing, is known as the primary figure in the development of modern probability-based time series analysis and forecasting, called Box–Jenkins methods.

Multivariate analysis was a prime interest of S. S. Wilks, who received his doctorate from the University of Iowa and followed it with post-graduate study at the University of London and at Cambridge University. Wilks then joined the Department of Mathematics at Princeton University. He worked continuously with the College Entrance Examining Board and the Educational Testing Service, both based in Princeton, New Jersey. During World

War II he directed the Statistical Research Group at Princeton in work performed for the Navy and the Army Air Force. [10]

Frank Yates succeeded R. A. Fisher as head of the Statistical Department at Rothamstead Experiment Station when Fisher accepted the Chair of Eugenics at University College, London in 1933. He remained in that post for some 40 years. His work in sampling is presented in *Sampling Methods for Censuses and Surveys* (Griffin, London, 1960). W. Edwards Deming advised the Department of Agriculture and the Bureau of the Census on sampling and surveys over a period of years. He produced a very useful book in 1960, *Sample Design in Business Research* (Wiley, New York). A major thrust in the book was the use of a method of replicated subsamples, which Deming acknowledges was first used by P. C. Mahalanobis is his 1936 surveys in Bengal. George Gallup was a true innovator in public opinion polling.

Mahalanobis founded the Indian Statistical Institute in 1931 and directed it for the rest of his life. Through his influence on Nehru, national sample surveys of india were developed in employment, agriculture, and industry. Hansen and his associates were the principal designers of large scale sample surveys for the Bureau of the Census in the United States in the 1940s and 1950s. In the early days, the development of sampling in India and the United States followed similar paths despite a lack of communication. [11]

Daniel Bell reports at length on a study by Karl Deutsch on social science advances between 1900 and 1965. Twelve of sixty-two achievements are represented quite directly by the fields of statistics and its near neighbors. These advances and the persons associated with them by Deutsch are [12].

5. Correlation analysis and social theory......................	Pearson, Edgeworth, Fisher
11. Intelligence tests...............	Binet, Terman, Spearman
27. Factor analysis..................	Thurstone
37. Large-scale sampling in social research...................	Hansen
39. National income accounting	Kuznets, C. Clark, U.N. Statistical Office
41. Attitude survey and opinion polling	Gallup, Cantril, Lazarsfeld. A. Campbell
46. Statistical decision theory	A. Wald

Near Neighbors

42. Imput–output analysis......	W. Leontief
47. Operations research and system analysis..................	Blackett, Morse, Bellman
51. Computers	Bush, Caldwell, Eckert, Mauchly
52. Multivariate analysis linked to social theory.......	Stouffer, T. W. Anderson, Lazarsfeld
54. Econometrics	Tinbergen, Samuelson, Malinvaud

Statistical methods have become an important technology in the modern world. We would hope with John Sinclair that they will be used to increase the "quantum of happiness" in the world.

We began with a request from Winston Churchill that brought out a distinction between *numbers* and *information*. The technology of the computer makes this distinction more important than ever. With Karl Pearson we would hope that statisticians deal in information that protects us from "quackery and dogma" and the appearance of knowledge where ignorance exists.

When it comes to the difference between numbers and information, Louis Armstrong put it as well as anyone.

A few well placed notes with love that swing are better than a thousand notes that don't say nothin'.

References

1. Churchill Eisenhart and Allan Birnbaum, "Aniversaries in 1966–67 of Interest to Statisticians," *The American Statistician*, **21**, No. 3 (June 1967), 22–29.
2. Maurice G. Kendall, "The History of Statistical Method," in *International Encyclopedia of Statistics* (W. H. Kruskal and J. M. Tanner, eds.), The Free Press, Glencoe, 1L, 1978, pp.1093–1101.
3. Edwin G. Boring, *A History of Experimental Psychology*, 2nd edn., Appleton-Century-Crofts, New York, 1950, pp. 482–483.
4. Paul A. Samuelson, "Harold Hoteling as Mathematical Economist," *The American Statistician*, **14** (June 1960), 21–25.
5. Maurice G. Kendall, "G. Udney Yule," in *International Encyclopedia of Statistics* (W. H. Kruskal and J. M. Tanner, eds.), The Free Press, Glencoe, IL, 1978, pp. 1261–1263.
6. Arthur P. Dempster and Frederick Mosteller, "In Memoriam—William Gemmell Cochran," *The American Statistician*, **35**, No. 1 (February 1981), 38.
7. Ralph A. Bradley, "Frank Wilcoxon," *The American Statistician*, **20** (February 1966), 32.
8. G. Udney Yule and M. G. Kendall, *An Introduction to the Theory of Statisitcs*, 14th edn., Hafner, New York, 1950.
9. Keith Ord, "In Memoriam—Maurice George Kendall," *The American Statistician*, **38**, No. 1 (February 1984), 36–37.
10. Frederick Mosteller, "Samuel S. Wilks: Statesman of Statistics," *The American Statistician*, **18** (April, 1964), 11–17.
11. W. Edwards Deming, "P. C. Mahalanobis," *The American Statistician*, **26**, No. 4 (October 1972), 49–50.
12. Daniel Bell, *The Social Sciences Since the Second World War*, Transaction Books, New Brunswick, NJ, 1982, pp. 12–23.

Bibliography

Aaron, Henry J., *Politics and the Professors: The Great Society in Perspective*, The Brookings Institute, Washington, DC, 1978.

Adams, Ernest W., "Survey of Bernoullian Utility," in *Mathematical Thinking in the Measurement of Behavior* (Herbert Solomon, ed.), The Free Press, Glencoe, Il, 1950.

Aitchison, John, *Choice Against Chance*, Addison-Wesley, Reading, MA, 1970.

American Statistical Association, *Acceptance Sampling*, Washington, DC, 1950.

Andrews, D. F., et al., *Robust Estimates of Location: Survey and Advances*, Princeton University Press, Princeton, NJ, 1972.

Ashurst, F. Gareth, *Pioneers of Computing*, Muller, London, 1983.

Bancroft, T. A., "George W. Snedecor, Pioneer Statistician," *The American Statistician*, **28**, No. 3 (August 1974), 108–109.

Barnard, G. A., "Thomas Bayes—A Biographical Note," in *Studies in the History of Statistics and Probability* (E. S. Pearson and M. G. Kendall, eds.), Griffin, London, 1970.

Bartlett, M. S., "R. A. Fisher and the Last 50 Years of Statistical Methodology," *Journal of the American Statistical Association*, **60**, No. 310 (June 1965), 395–409.

Bayes, Thomas, "An Essay Towards Solving a Problem in the Doctrine of Chances," in (E. S. Pearson and M. C. Kendall, eds.) *Studies in the History of Statistics and Probability*, Griffin, London, 1970.

Bell, Daniel, *The Social Sciences Since the Second World War*, Transaction Books, New Brunswick, NJ, 1982.

Bernoulli, D., "Exposition on a New Theory on the Measurement of Risk," (L. Sommer, transl.), *Econometrika*, **22** (1954), 23–36.

Boneau, C. A., "The Effects of Violations of Assumptions Underlying the *t*-Test," *Psychological Bulletin*, **57** (1960), 49–64.

Boring, Emil G., *A History of Experimental Psychology*, 2nd edn., Appleton-Century-Crofts, New York, 1950.

Bowler, M. Kenneth, *The Nixon Guaranteed Income Proposal*, Ballinger, Cambridge, MA, 1974.

Bowley, A. L., *An Introduction to the Theory of Statistics*, P. S. King & Son, London, 1920.

Box, G. E. P., "Robustness in the Strategy of Scientific Model Building," in *Robustness in Statistics* (Robert W. Launer and Graham N. Wilkinson, eds.), Academic Press, New York, 1979, pp. 201–236.

Box, Joan Fisher, *R. A. Fisher: The Life of a Scientist*, Wiley, New York, 1978.

Box, Joan Fisher, "Gosset, Fisher, and the *t*-Distribution," *The American Statistician*, **35**, No. 2 (May 1981), 61–66.

Bradley, Ralph A., "Frank Wilcoxon," *The American Statistician*, **22** (February 1966), 32.

Bryant, Edward C., "Survey Statistics in Social Program Evaluation," in *Papers in Honor of H. O. Hartley*, (H. A. David, ed.), Academic Press, New York, 1978.

Bureau of Labor Statistics, *The Consumer Price Index: Concepts and Content Over the Years*, BLS Report 517, May 1978 (Revised).

Burt, Cryril, "Francis Galton and His Contributions to Psychology," *British Journal of Statistical Psychology*, **13** (May, 1962), 31.

Campbell, Donald T., and Robert F. Boruch, "Making the Case for Randomized Assignment to Treatments by Considering the Alternatives: Six Ways in Which Quasi-Experimental Evaluations in Compensatory Education Tend to Underestimate Effects," in *Evaluation and Experiment* (Carl A. Bennett and Arthur A. Lumsdaine, eds.) Academic Press, New York, 1975, pp. 195–275.

Campbell, Donald T., and Julian C. Stanley, *Experimental and Quasi-Experimental Designs for Research*, Rand McNally, Chicago, 1966.

Campbell, Kathy, and Brian L. Joiner, "How to Get the Answer Without Being Sure You've Asked the Question," *The American Statistician*, **27**, No. 5 (December 1973), 229–231.

Carson, Carol S., "The History of the United States National Income and Product Accounts," *Review of Income and Wealth*, **Ser. 21** (June 1975), 153–181.

Cassedy, James, *Demography in Early America*, Harvard University Press, Cambridge, MA, 1969.

Cavicehi, R. E., letter to *The American Statistician*, **26**, No. 4 (October 1972), 25.

Chernoff, H., and L. E. Moses, *Elementary Decision Theory*, Wiley, New York, 1959.

Churchill, Winston, *The Gathering Storm*, Houghton Mifflin, New York, 1948.

Churchman, C. West, *Theory of Experimental Inference*, Macmillan, New York, 1948.

Clague, Ewan, *The Bureau of Labor Statistics*, Praeger, New York, 1968.

Cochran, W. G., F. Mosteller, and J. W. Tukey, "Principles of Sampling," *Journal of the American Statistical Association*, **49**, No. 265 (March 1954), 13–35.

Collins, Sharon M., "The Use of Social Research in the Courts," in *Knowledge and Policy: The Uncertain Connection*, National Research Council Study Project on Social Research and Development, National Academy of Sciences, Washington, DC, 1978.

Colton, Theodore, and Robert C. Buxbaum, "Motor Vehicle Inspection and Motor Vehicle Accident Mortality," in *Statistics and Public Policy* (William B. Fairley and Frederick Mosteller, eds.), Addison-Wesley, Reading, MA, 1977, pp. 131–142.

Cook, Thomas D., and Donald T. Campbell, *Quasi-Experimentation: Design and Analysis Issues for Field Settings*, Houghton Mifflin, Boston, 1979.

Cooley, William W., and Paul R. Lohnes, *Evaluation Research in Education*, Irvington, New York, 1976.

David, F. N., *Games, Gods, and Gambling*, Hafner, New York, 1962.

Davidson, D., S. Seigel, and P. Suppes, *Decision Making*, Stanford University Press, Stanford, CA, 1957.

Day, B. B., "The Design of Experiments in Naval Engineering," *The American Statistician*, **8** (April–May 1954), 7–12, 23.

Deming, W. Edwards, *Elementary Principle of Statistical Control of Quality*, rev. 2nd edn., Nippon Kagaku Gijutsu Remmei, Tokyo, 1952.

Deming, W. Edwards, *Sample Design in Business Research*, Wiley, New York, 1960.

Deming, W. Edwards, "P. C. Mahalanobis," *The American Statistician*, **26**, No. 4 (October 1972), 49–50.

Deloitte, Haskins, and Sells, *The Week in Review*, **80–10** (March 7, 1980).

Dempster, Arthur P., and Frederick Mosteller, "In Memoriam—William G. Cochran, *The American Statistician*, **35**, No. 1 (February 1981) 1, 38.

Dudding, Bernard P., "The Introduction of Statistical Methods to Industry," *Applied Statistics*, **1** (1952), 3–20.

Duncan, Acheson, J., *Quality Control and Industrial Statistics*, 4th edn., Irwin, Homewood, II, 1974.

Efrom, Bradley, and Gail Gong, "A Liesurely Look at the Bootstrap, the Jacknife, and Cross-validation," *The American Statistician*, **37**, No. 1 (February 1983), 36–48.

Eisenhart, Churchill, "On The Transition from Student's *z* to Student's *t*," *The American Statistician*, **33**, No. 1 (February 1979), 8–10.

Eisenhart, Churchill, and Allan Birnbaum, "Anniversaries in 1966–67 of Interest to Statisticians," *The American Statistician*, **21**, No. 3 (June 1967), 22–29.

Evans, Brian, and Bernard Waites, *IQ and Mental Testing*, Macmillan, New York, 1981.

Ezekiel, Mordecai, J. B., and Karl A. Fox, *Correlation and Regression Analysis*, Wiley, New York, 1959.

Fellner, William, *Probability and Profit*, Irwin, Homewood, II, 1965.

Ferber, Robert, and Werner Z. Hirsch, *Social Experimentation and Economic Policy*, Cambridge University Press, Cambridge, UK, 1982.

Fishbein, Meyer H., "The Censuses of Manufactures, 1810–1890," *National Archives Accessions*, No. 57 (June 1963), 20.

Fisher, Arne, *The Mathematical Theory of Probabilities*, Macmillan, New York, 1923.

Fisher, Irving, *The Making of Index Numbers*, Houghton Mifflin, Boston, 1923.

Fisher, R. A., "The Use of Multiple Measurements in Taxonomic Problems," *Annals of Eugenics*, **7** (1936), 179–188.

Fisher, R. A., *The Design of Experiments*, 3rd edn., Oliver and Boyd, London, 1942.

Fitzpatrick, Paul J., "Leading American Statisticians in the Nineteenth Century," *Journal of the American Statistical Association*, **52**, No. 279 (September 1957), 301–321.

Fitzpartrick, Paul J., "Leading British Statisticians of the Nineteenth Century," *Journal of the American Statistical Association*, **55**, No. 289 (March 1960).

Fortune, "Dr. Deming Shows Pontiac the Way," April 18, 1983, p. 66.

Friedman, Milton, and L. J. Savage, "The Utility Analysis of Choices Involving Risk," *Journal of Political Economy*, **56** (1948), 279–304.

Gaito, John, "Non-parametric Methods in Psychological Research," *Psychological Reports*, **5** (1959), 115–125.

Gallup, George, *The Sophisticated Pollwatcher's Guide*, Princetion University Press, Princeton, NJ, 1972.

Gallup Organization, Inc., The, *Design of the Gallup Sample*, Princeton, NJ,

Galton, Francis, "Classification of Men According to Their Natural Gifts," in *The World of Mathematics* (James R. Newman, ed.), Simon and Schuster, New York, 1956, pp. 1173–1191.

Galton, Francis, *Memories of My Life*, Metheun, London, 1908.

Garvin, David A., "Quality on the Line," *Harvard Business Review*, **61** (September–October 1983), 65–75.

Good, I. J., *Good Thinking: The Foundations of Probability and Its Applications*, University of Minnesota Press, Minncapolis, MN, 1983.

Goodall, Colin, "M-Estimators of Location: An Outline of the Theory," in *Understanding Robust and Exploratory Data Analysis* (David C. Hoaglin, Frederick Mosteller, and John W. Tukey, eds.), Wiley, New York, 1983, pp. 339–403.

Gosset, William S., "The Probable Error of a Mean," *Biometrika*, **6** (1908), 1–25.

Gosset, William S., "Errors of Routine Analysis," *Biometrika*, **19** (1927), 151–164.

Grant, Eugene L., "Industrialists and Professors in Quality Control," *Industrial Quality Control*, **X**, No. 1 (July 1953), 31–35.

Grayson, C. Jackson, Jr., *Decisions Under Uncertainty: Drilling Decisions by Oil and Gas Operators*, Graduate School of Business Administration, Harvard University, Cambridge, MA, 1960.

Green Paul E., "Risk Attitudes and Chemical Investment Decisions," *Chemical Engineering Progress*, **59**, No. 1 (January 1963), 35–40.

Griffin, J., *Statistics: Methods and Applications*, Holt, Rinehart, and Winston, New York, 1962.

Griffith, R. M., "Odds Adjustment by by American Horse-Race Bettors," *American Journal of Psychology*, **62** (1949), 290–294.

Gryna, Frank M., *Quality Circles: A Team Approach to Problem Solving*, American Management Association, New York, 1981.

Gulland, J. A., *Manual of Sampling and Statistical Methods for Fisheries Biology*, United Nations, New York, 1966.

Hankins, Frank H., *Adoplhe Quetelet as Statistician*, Columbia University Press, New York, 1908.

Hartwig, Frederick, and Brian E. Dearing, *Exploratory Data Analysis*, Addison-Wesley, Reading, MA, 1977.

Hays, William L., and Robert Winkler, *Statistics: Probability, Inference, and Decision*, Holt, Rinehart, and Winston, New York, 1970.

Heermann, Emil F., and Larry A. Braskamp, *Readings in Statistics for the Behavioral Sciences*, Prentice-Hall, Englewood Cliffs, NJ, 1970.

Hogben, Lancelot, *Chance and Choice by Cardpack and Chessboard*, Parrish, London, 1955.

Hopkins, Harry, *The Numbers Game*, Little, Brown, Boston, 1973.

Hotelling, Harold, "Analysis of a Complex of Statistical Variables into Principal Components," *Journal of Educational Psychology*, **24** (1933), 417–441 and 498–528.

Hotelling, Harold, "The Impact of R. A. Fisher on Statistics," *Journal of the American Statistical Association*, **46**, No. 253 (March 1951), 37.

Huber, George, "Methods for Quantifying Subjective Probabilities," *Decision Sciences*, **5**, No. 3 (July 1974), 3–31.

Janowitz, M., R. Lekachman, D. P. Moynihan, *et al.*, "Social Science: The Public Disenchantment: A Symposium," *American Scholar*, **45** (1976), 335–359.

Jaszi, George, "Bureau of Economic Analysis," in *Encyclopedia of Economics* (Douglas Greenwald, ed.), McGraw-Hill, New York, 1982.

Juran, J. M., "Japanese and Western Quality—A Contrast," *Quality Progress*, **XI** (December 1978), 18.

Kendall, Maurice G., "The History of Statistical Method," in *International Encyclopedia of Statistics* (W. H. Kruskal and J. M. Tanner, eds.), The Free Press, Clencol, IL, 1978, pp. 1093–1101.

Kendall, Maurice G., "G. Udney Yule", *International Encyclopedia of Statistics* (W. H. Kruskal and J. M. Tanner, eds.), The Free Press, Glencoe, IL, 1978, pp. 1261–1263.

Kevles, Daniel J., "Annals of Eugenics, Parts I–IV," *The New Yorker*, October 8, 15, 22, 29, 1984.

Keynes, John Maynard, *Essays in Biography*, Macmillan, London, 1933.

Kirk, Roger E., *Statistical Issues: A Reader for the Behavioral Sciences*, Wadsworth, Belmont, CA, 1972.

Kish, Leslie, "Some Statistical Problems in Research Design," *American Sociological Review*, **24** (June 1959), 328–338.

Kruskal, William, and Frederick Mosteller, "Representative Sampling, IV: The History of the Concept in Statistics, 1895–1939," *International Statistical Review*, **48** (1980), 169–195.

Langbein, Laura Irwin, *Discovering Whether Programs Work: A Guide to Statistical Methods for Program Evaluation*, Scott, Foresman, Glencoe, II, 1980.

Laplace, Pierre Simon, *A Philosophical Essay on Probabilities*, (translated from the sixth French edition), Dover, New York, 1951.

Lehman, E. H., *Nonparametrics*, Holden-Day, San Francisco, 1975.

Leonard, William R., "Walter Willcox: Statist," *The American Statistician*, **15** (February 1961), 16–19.

Levine, Robert A., "How and Why the Experiment Came About," in *Work Incentives and Income Guarantees: The New Jersey Negative Income Tax Experiment*, (Joseph A. Pechman and Michael A. Timpane, eds.), The Brookings Institute, Washington, DC, 1975.

Lichtenstein, Sarah, *et al.*, "Calibration of Probabilities: The State of the Art," in *Decision Making and Change in Human Affairs*, (H. Jungermann and G. de Zeeuw, eds.), Reidel, Dordrecht, Holland, 1977, pp. 275–324.

Leiby, James, *Carol Wright and Labor Reform: The Origin of Labor Statistics*, Harvard University Press, Cambridge, MA, 1960.

Levitan, Sar A., and Gregory K. Wurzburg, *Evaluating Federal Programs*, W. E. Upjohn Institute, Kalamazoo, MI, 1979.

Lindley, D. V., *Introduction to Probability and Statistics from a Bayesian Viewpoint*, Cambridge University Press, Cambridge, UK, 1965.

Littauer, S. B., "The Development of Statistical Quality Control in the United States," *The American Statistician*, **4** (December 1950), 14–20.

Lush, Jay L., "Early Statistics at Iowa State University," in *Statistical Papers in Honor of George W. Snedecor*, (T. A. Bancroft, ed.) Iowa State University Press, Ames, IA, 1972, pp. 211–226.

Main, Jeremy, "The Curmudgeon Who Talks Tough on Quality," *Fortune*, June 25, 1984, 118–122.

Markham, Jesse W. and Paul V. Teplitz, "*Baseball Economics and Public Policy*, Lexington Books, Lexington, MA, 1981.

McKinley, Sonja M., "The Design and Analysis of the Observational Study—A Review," *Journal of the American Statistical Association*, **70**, No. 351 (September 1975), 503–520.

McMullen, Launce, "Student as a Man," *Biometrika*, **30** (1939), 205–210.

Meier, Paul, "The Biggest Public Health Experiment Ever: The 1954 Field Trial of the Salk Poliomyelitis Vaccine," in *Statistics: A Guide to the Unknown* (Judith Tanner, ed.), Holden-Day, San Francisco, 1972, pp. 2–13.

Miller, George A. (ed.), *Mathematics and Psychology*, Wiley, New York, 1964.

Moore, Henry L., *Economic Cycles, Their Law and Cause*, Macmillan, London, 1914.

Moroney, M. J., *Facts from Figures*, 3rd ed., Penguin, Baltimore, MD, 1956.

Morris, Charles, *Varieties of Human Value*, University of Chicago Press, Chicago, 1956.

Morris, Charles, and Linwood Small, "Changes in the Conception of the Good Life by American College Students from 1950 to 1970," *Journal of Personality and Social Psychology*, **20**, No. 2 (1971), 254–260.

Mosteller, F. *et al.*, *The Pre-election Polls of 1948*, Social Science Research Council (Bulletin 60), New York, 1949.

Mosteller, F., and P. Nogee, "An Experimental Measurement of Utility," *Journal of Political Economy*, **59** (1951), 371–404.

Mosteller, Frederick, and D. L. Wallace, *Inference and Disputed Authorship: The Federalist*, Addison-Wesley, Reading, MA, 1964.

Mosteller, Frederick, "Samuel S. Wilks: Statesman of Statistics," *The American Statistician*, **18** (April 1964), 11–17.

Murphy, Allan H., and Robert L. Winkler, "Reliability of Subjective Probability Forecasts of Precipitation and Temperature," *Applied Statistics*, **26**, No. 1 (1977), 41–47.

Newsweek Magazine, "Huh," November 39, 1959.

Neyman, Jerzy, "R. A. Fisher: An Appreciation," *Science*, **156** (1967), 1456–1460.

Norwood, Janet, "Unemployment and Associated Measures," in *The Handbook of Economic and Financial Measures* (Frank J. Fabozzi and Harry I Greenfield, eds.), Dow Jones–Irwin, Homewood, IL, 1984.

Norwood, Janet L., and John F. Early, "A Century of Methodological Progress at the U.S. Bureau of Labor Statistics," *Journal of the American Statistical Association*, **79**, No. 388 (December 1984), 748–761.

Ord, Keith, "In Memoriam—Maurice G. Kendall," *The American Statistician*, **38**, No. 1 (February 1984), 36–37.

Osgood, Charles, G. J. Suci, and P. H. Tannenbaum, *The Measurement of Meaning*, University of Illinois Press, Urbana, II, 1957.

Osgood, Charles, Edward E. Ware, and Charles Morris, "Analysis of Connotative Meanings of a Variety of Human Values as Expressed by College Students," *Journal of Abnormal and Social Psychology*, **62**, No. 1 (1961), 62–73.

Parade Magazine, "Ranking of Presidents by American Historians," February 6, 1984.

Pearson, E. S., *Karl Pearson: An Appreciation of Some Aspects of His Life and Work*, Cambridge University Press, Cambridge, UK, 1938.

Pearson, E. S., "Student as a Statistician," *Biometrika*, **30** (1939), 210–250.

Pearson, E. S., and M. G. Kendall (eds.), *Studies in the History of Statistics and Probability*, Griffin, London, 1970.

Pearson, Karl, *The Grammar of Science*, Dent, London, 1937.

Peters, William S., *Readings in Applied Statistics*, Prentice-Hall, Englewood Cliffs, NJ, 1969.

Preston, M. G., and P. Barratra, "An Experimental Study of the Auction Value of an Uncertain Outcome," *American Journal of Psychology*, **61** (1948), 183–193.

Quenouille, M. H., "Notes on Bias in Estimation," *Biometrika*, **43** (1956), 353–360.

Raiffa, Howard, *Decision Analysis: Introductory Lectures on Choices Under Uncertainty*, Addison-Wesley, Reading, MA, 1968.

Ramsey, Frank P., *The Foundations of Mathematics*, Harcourt Brace, New York, 1931.

Reid, Constance, *Neyman—from Life*, Springer-Verlag, New York, 1982.

Rickey, Branch, "Goodbye to Some Old Baseball Ideas," *Life Magazine*, August 2, 1954.

Rombauer, Irma S., *The Joy of Cooking*, Blakiston, Philadelphia, PA, 1943.

Roos, C. F., and V. von Szeliski, *The Dynamics of Automobile Demand*, General Motors, New York, 1939.

Rosenberger, James L., and Miriam Gasko, "Comparing Location Estimators: Trimmed Means, Medians, and Trimean," in *Understanding Robust and Exploratory Data Analysis* (David C. Hoaglin, Frederick Mosteller, and John W. Tukey, eds.), Wiley, New York, 1983, pp. 297–337.

Rubin, Donald R., "William G. Cochran's Contributions to the Design, Analysis, and Evaluation of Observational Studies," in *W. G. Cochran's Impact on Statistics* (Poduri S. Rao and Joseph Sedransk, eds.), Wiley, New York, 1984, pp. 37–69.

Sampson, A. R., Letter to *The American Statistician*, **28**, No. 2 (May 1974), 76.

Savage, L. J., "Subjective Probability and Statistical Practice," in *The Foundations of Statistical Inference* (M. S. Bartlett, ed.), Methuen, London, 1962, pp. 9–35.

Schlaifer, Robert, *Probability and Statistics for Business Decisions: An Introduction to Managerial Economics Under Uncertainty*, McGraw-Hill, New York, 1959.

Scully, Gerald W., "Pay and Performance in Major League Baseball," *American Economic Review*, **64**, No. 6 (December 1964), 915–930.

Shearer, Lloyd, "Pierpoint's Presidential Report Card," *Parade Magazine*, January 7, 1982.

Shurkin, Joel, *Engines of the Mind: A History of the Computer*, Norton, New York, 1984.

Simpson, E. H., "The Interpretation of Interaction in Contingency Tables," *Journal of the Royal Statistical Society*, Ser. B, **13** (1951), 238–241.

Singleman, J. *From Agriculture to Service*, Sage Publications, Beverly Hills, 1978.

Sinkey, Joseph F., Jr., *Problem and Failed Institutions in the Commercial Banking Industry*, JAI Press, Greenwich, CT, 1979.

Slonim, Morris J., "Sampling in a Nutshell," *Journal of the American Statistical Association*, **52**, No. 278 (June 1957), 143–161.

Slonim, Morris J., *Sampling in a Nutshell*, Simon and Schuster, New York, 1960.

Solomon, Herbert, "A Survey of Mathematical Models in Factor Analysis," in *Mathematical Thinking in the Measurement of Behavior* (Herbert Solomon, ed.), The Free Press, Glencoe, IL, 1960, pp. 273–313.

Spearman, Charles, "General Intelligence, "*American Journal of Psychology*, **15** (1904), 232.

Speigelman, Robert G., and K. E. Yeager, "Overview," *The Journal of Human Resources*, **XV**, 4 (1980), 463–479.

Street, Elisabeth, and Mavis B. Carroll, "Preliminary Evaluation of a Food Product," in *Statistics: A Guide to the Unknown* (Judith M. Tanur, ed.), Holden-Day, San Francisco, 1972, pp. 220–228.

Stigler, Stephen M., "Eight Centuries of Sampling Inspection: The Trial of the Pyx," *Journal of the American Statistical Association*, **72**, No. 359 (September 1977), 493–500.

Stigler, Stephen M., "Do Robust Estimators Work with Real Data?" *The Annals of Statistics*, **5**, No. 6 (1977), 1055–1098.

Tankard, James W., *The Statistical Pioneers*, Schenkman, Cambridge, MA, 1984.

Thomson, Godfrey H., *The Factorial Analysis of Human Ability*, Houghton Mifflin, Boston, 1939.

Thurstone, Louis L., *The Measurement of Values*, University of Chicago Press, Chicago, 1959.

Tintner, Gerhard, *Econometrics*, Wiley, New York, 1952.

Tippett, L. H. C., *The Methods of Statistics*, 4th edn., Wiley, New York, 1952.

Tippett, L. H. C., "The Making of an Industrial Statistician," in *The Making of Statisticians*, (J. Gani, ed.), Springer-Verlag, New York, 1982, pp. 182–187.

Tufte, Edward R., *Data Analysis for Politics and Policy*, Prentice-Hall, Engelwood Cliffs, NJ, 1974.

Tukey, John W., *Exploratory Data Analysis*, Addison-Wesley, Reading, MA, 1977.

U.S. Bureau of the Census, "Factfinder for the Nation," CFF No 4 (Rev), May, 1979.

Vellemen, Paul F., and David C. Hoaglin, *Applications, Basics, and Computing of Exploratory Data Analysis*, Duxbury Press, Boston, MA, 1981.

Verhoeven, C. J., *Techniques in Corporate Planning*, Kluher Nijoff, Amsterdam, 1982.

Vincent, Douglas, The Origin and Development of Factor Analysis," *Applied Statistics*, **2** (1953), 116.

Von Mises, Richard, *Probability, Statistics, and Truth*, Allen and Unwin, London, 1939.

Von Neumann, J., and O. Morgenstern, *Theory of Games and Economic Behavior*, 2nd edn., Princeton University Press, Princeton, NJ, 1947.

Wagner, Clifford H., "Simpson's Paradox in Real Life," *The American Statistician*, **36**, No. 1 (February 1982), 48.

Walker, Helen M., *Studies in the History of Statistical Method*, Williams and Wilkins, Baltimore, MD, 1929.

Walker, Helen M., *Elementary Statistical Methods*, Holt, New York, 1943.

Walker, Helen M., "The Contributions of Karl Pearson," *Journal of the American Statistical Association*, 53, No. 281 (March 1958), 11–22.

Wallis, W. Allen, "The Statistical Research Group, 1942–1945," *Journal of the American Statistical Association*, 75, No. 370 (June 1980), 320–330.

Weaver, Warren, *Lady Luck*, Anchor Books, New York, 1963.

Wells, Oris V., Mordecai J. B. Ezekiel, 1899–1974," *The American Statistician*, 29, No. 2 (May, 1975), 106.

Wheeler, Michael, *Lies, Damn Lies, and Statistics*, Liveright, New York, 1976.

Wilks, S. S., "Karl Pearson: Founder of the Science of Statistics," *The Scientific Monthly*, 53 (1941), 249–253.

Williams, Bill, *A Sampler on Sampling*, Wiley, New York, 1978.

Working, E. J., "What Do Statistical 'Demand Curves' Show," *Quarterly Journal of Economics*, 4 (1927), 212–235.

Working, Holbrook, "Statistical Quality Control in War Production," *Journal of the American Statistical Association*, 40, No. 2 (December 1945), 425–447.

Wright, Sewell, "Correlation and Causation," *Journal of Agricultural Research*, 20 (1921), 557–585.

Wright, Sewell, "The Method of Path Coefficients, *Annals of Mathematical Statistics*, 5 (1934), 161–215.

Wright, Sewell, *Evolution and the Genetics of Populations, Vol I, Genetic and Biometric Foundations*, University of Chicago Press, Chicago, 1968.

Yates, F., "Sir Ronald Fisher and the Design of Experiments," *Biometrics*, 20 (June 1964), 307–315.

Youden, W. J., "How to Pick a Winner," *Industrial and Engineering Chemistry*, 50, No. 6 (1958), 81A.

Youden, W. J., "Random Numbers Aren't Nonsense," *Industrial and Engineering Chemistry*, 49, No. 10 (October 1957), 89A.

Youden, W. J., "Factorial Experiments Help Improve Your Food," *Industrial and Engineering Chemistry*, 49, No. 2 (February 1957), 85A.

Youden, W. J., *Experimentation and Measurement*, National Bureau of Standards Special Publication 672, 1984.

Yule, G. Udney, *An Introduction to the Theory of Statistics*, 5th edn., Griffin, London, 1922.

Yule, G. Udney, and M. G. Kendall, *An Introduction to the Theory of Statistics*, Hafner, New York, 1950.

List of Names

Subject Index

DATE DUE			
MAY 1 7 '90			
5/20/90			

FRANKLIN PIERCE COLLEGE

LIBRARY

Rindge, NH   03461